Solar Power

Solar Power

by Ally Stone (DeGunther)
Previous edition by Rik DeGunther

Solar Power For Dummies®

Contents at a Glance

Table of Contents

Introduction

Solar power is no longer a niche technology or a futuristic idea. It's a rapidly growing part of how the world produces and uses energy. From rooftop systems on homes to massive solar farms powering entire regions, solar energy is transforming electricity generation at every scale. At the same time, falling costs, financing options, advances in storage, and smarter software tools have made solar more accessible than ever.

This book guides you through that evolving landscape. Whether you're a homeowner exploring your first installation, a business leader evaluating energy options, or simply curious about how solar power works and where it's headed, you'll find practical explanations without unnecessary jargon. You don't need an engineering degree to understand solar power, just curiosity and a willingness to explore.

About This Book

Solar Power For Dummies explains how solar energy works, how it's used in homes and businesses, and how it fits into the larger power grid. Unlike many resources that focus only on residential installations, this book covers solar across all scales: from small rooftop systems to utility-scale projects and emerging technologies such as space-based solar and advanced storage.

You'll find clear explanations, real-world examples, and guidance to help you make informed decisions, whether that means installing solar, investing in it, or simply understanding its role in the energy transition. Like all For Dummies books, this one is organized for easy access, so you can jump to the sections most relevant to you without reading cover to cover.

To make the content more accessible, I divided it into six parts:

>> **Part 1: Here Comes the Sun.** This part explains how solar energy woks and why it matters today.

>> **Part 2: Investing in Residential Solar Power: Powering Your Home.** This part guides homeowners through establishing home energy efficiency, and installing, financing, and living with a solar power system.

>> **Part 3: Examining Commercial and Industrial Solar Projects.** In this part, you can find out how businesses evaluate, finance, and deploy solar power projects.

>> **Part 4: Utility-Scale Solar: Powering the Grid.** This part explores large solar projects — including their development and financing — as well as the scope of power markets.

>> **Part 5: Seeing What's Next for Solar: The Future Is Bright.** This part looks ahead to energy storage, advanced solar cells, software, and space-based solar power.

>> **Part 6: The Part of Tens.** This fun Dummies part provides quick lists, tips, benefits, misconceptions, and practical takeaways regarding solar power.

Foolish Assumptions

I assume that you're interested in solar power but may have little or no technical background. You might be a homeowner considering panels, a business professional evaluating energy costs, a student, or simply someone curious about clean energy.

At the same time, I recognize that readers come with different levels of experience. Some sections start with the basics, while others explore deeper topics such as financing structures, grid integration, or emerging technologies. Feel free to skip around in the book to the material that matches your interests and comfort level.

Icons Used in This Book

Throughout this book, icons in the margins highlight certain types of valuable information that call out for your attention. Here are the icons you'll encounter and a brief description of each.

The Tip icon marks tips and shortcuts that will help you get the most out of solar power and energy decisions.

Remember icons mark the information that's especially important to know. To siphon off the most important information in each chapter, just skim through these icons.

The Technical Stuff icon marks information of a highly technical nature that you can normally skip over.

The Warning icon tells you to watch out! It marks important information that may save you headaches, or point out safety issues, financial risks, or common mistakes that would cost you time or money.

Beyond the Book

In addition to the abundance of information and guidance related to Solar Power that we provide in this book, you get access to even more help and information online at Dummies.com. Check out this book's online Cheat Sheet. Just go to www.dummies.com and type "Solar Power For Dummies Cheat Sheet" into the Search box.

Where to Go from Here

This book isn't meant to be read strictly from start to finish. If you're new to the topic of solar power, start with Part 1. Homeowners thinking about installing a solar power system may want to jump to Part 2, and business readers can head straight to Part 3. If you're curious about the future of solar energy, Part 5 explores what's coming next.

Wherever you begin, you'll gain a clearer picture of how solar power works — and how it's reshaping the way people produce and use energy.

1
Here Comes the Sun

Explore how solar power fits into the global energy transition — offering affordability, versatility, and independence from fossil fuels — and find out why it's often called the king of renewables.

Break down typical household energy use, compare it to national trends, and analyze your own electricity bill so you know exactly what you're paying for and where solar can make the biggest difference.

Get a crash course in the Sun itself by charting its seasonal intensity and grasping its makeup — from nuclear fusion at its core to the photons that deliver energy to Earth.

Chapter **1**

Solar Power in Today's Energy Transition

Welcome to the bright side of the energy transition! In this chapter, you get an up-to-date snapshot of how solar power has gone from a niche option to the superstar of today's sustainable energy movement. I walk you through the reasons that solar power stands out among renewable energy forms — from its ability to slash carbon emissions to its status as the most affordable electricity in history.

You also explore how solar power fits projects of every size: from rooftop systems powering single homes to vast solar farms producing energy for entire cities. And because no energy solution is perfect, I tell you upfront about the challenges of collecting and using solar power, which include timing of power production, reliability when several cloudy days can deplete stored energy, and regulatory hurdles. The chapter content helps you understand why solar power is leading the charge, where it shines brightest, and what to watch out for as systems shift toward a cleaner energy future.

Looking for Sustainable Energy

If you've been paying even a little attention to the news media, you know the world is in the middle of a massive energy transformation in response to rising greenhouse gas emissions, volatile fossil fuel prices, and the urgent need to address climate change. These issues are pushing governments, businesses, and individuals to seek more sustainable and affordable ways to power our future.

Sustainable energy isn't just a buzzword; it's a real mission that involves producing power without depleting natural resources or causing long-term environmental harm. The goal is simple: meet today's energy needs while building a long-term, stable energy future. Solar energy checks every box in this mission, making it a natural starting point for anyone exploring clean energy solutions.

When you hear the words *sustainable energy*, think of sources that are renewable (can be replenished naturally) and have minimal environmental impact — think solar, wind, and hydro power.

Understanding Why Solar Is King

Solar power's dominance isn't just about being renewable, it's the combination of accessibility, scalability, and long-term value that solar power solutions deliver. Over the past two decades, the technology has matured, costs have plummeted, and adoption has spread from early adopters to mainstream homeowners, corporations, and utilities.

Solar power is not just an alternative: In many places, it's the preferred choice for new power generation. Solar has taken center stage because its universal appeal offers something for everyone:

>> Environmental gains for those who care about climate

>> Economic benefits for cost-conscious users

>> Energy security for those focused on reliability

Erasing your carbon footprint

While motivations for going solar vary, one of the most tangible benefits is the ability to dramatically cut greenhouse gas emissions. A typical residential solar system can offset several tons of carbon dioxide (CO_2) per year, which is equivalent to planting dozens of trees annually. For business and utilities, the impact is magnified, offsetting tens or even hundreds of thousands of tons of CO_2 each year. Table 1-1 shows you the CO_2 offset for types of solar installations.

TABLE 1-1   **Annual CO_2 Offset per Type of Solar Installation**

System Type	Annual CO_2 Offset	Equivalent Cars Removed from Road	Equivalent Trees Planted
Residential	3–4 metric tons	0.8	50–67
Commercial	300–450 metric tons	65–98	5,000–7,500
Utility-Scale	85,000–100,000 metric tons	18,500–21,000	1.4–1.7 M

U.S. Environmental Protection Agency/https://www.epa.gov/energy/greenhouse-gas-equivalencies-calculator/Public domain/last accessed Mar. 4, 2026.

When I talk about *offsetting emissions*, I mean reducing or eliminating the amount of pollution that would otherwise be created. In practical terms every kilowatt-hour (kWh) of electricity generated by solar is one less kilowatt-hour that needs to come from fossil fuel power plants. This displacement prevents those fossil fuel plants from burning fuels and releasing CO_2, effectively keeping that pollution out of the atmosphere.

REMEMBER

Beyond broader climate considerations, these reductions in emissions improve local air quality, leading to measurable health benefits. Reduced reliance on coal- and gas-fired power plants means fewer particulates, nitrogen oxides, and sulfur dioxide in the air. These pollutants are linked to asthma, heart disease, and other serious health conditions. Cleaner air is better for the community.

Enjoying solar's unlimited supply

Unlike fossil fuels, which are finite and subject to depletion, the Sun provides steady, predictable energy source that cannot be monopolized, embargoed, or depleted. Every hour, the sunlight hitting the Earth contains more energy than the entire human population consumes in a year. Harnessing even a fraction of that energy can meet the world's electricity needs many times over.

The abundance of sunlight energy creates strategic advantages. Nations with limited fossil fuel reserves can use solar to bypass the need to import costly fuels and, instead, build domestic energy independence. In rural or off-grid communities, solar energy can provide electricity where extending transmission lines is too expensive or impractical. On a personal scale, having access to the Sun's energy means that you can produce electricity without worrying about depletion or sudden shortages.

Long-term, this unlimited supply gives solar energy a resilience other resources can't match. Whereas fossil fuel reserves require constant exploration, drilling, and transport, the Sun rises every morning without human intervention.

Embracing solar as the most affordable form of energy

Solar power has undergone one of the most dramatic cost reductions in the history of energy. A big part of the overall cost of solar is the cost of the solar panels themselves, which is, on average, 20 percent of the costs. From 2010 to 2020, the cost of solar panels dropped by more than 80 percent, driven by advances in manufacturing, economies of scale, and improved efficiency. Installation methods have become faster and less labor-intensive, while financial models (in which costs are spread overtime through loans, leases, or power purchase agreements) have made solar accessible to even more people.

In many markets globally, solar is now the cheapest form of new electricity generation, even without government incentives. This cost advantage extends from residential rooftops to massive solar farms. For homeowners, this can mean locking in low electricity rates for many years and protecting against future price spikes in traditional power. For businesses, the savings can be substantial enough to boost profitability, reinvest in operations, or pass cost reductions to customers. Figure 1-1 shows a chart published on the Taiyang News website (`https://taiyangnews.info/markets/us-solar-installations-h1-2025`) that illustrates the price drop and increasing in installations in the U.S. since 2011.

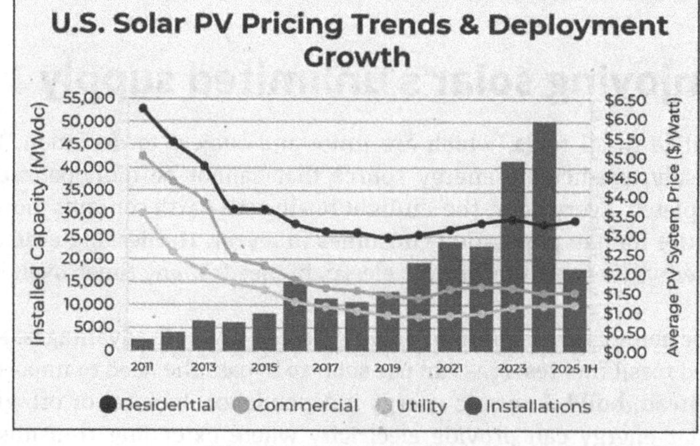

FIGURE 1-1: Solar power installations have increased, and the cost has decreased since 2011.

TaiyangNews/US Installed 18 GW New Solar PV Capacity in H1 2025/ https://taiyangnews.info/markets/us-solar-installations-h1-2025/last accessed Mar. 4, 2026

Even if you're not ready to install solar panels yourself, many utilities now offer community solar programs that enable you to subscribe to solar power and enjoy cost savings without owning or maintaining the panels.

Appreciating solar energy's versatility

Solar power systems aren't a one-size-fits-all technology, and that's part of their strength. Solar power users can install panels on residential rooftops, integrate collectors into building facades, or deploy a system of panels as ground-mounted arrays in open fields. They can power homes in dense urban areas, supply electricity to manufacturing facilities, or bring light and communication to rural communities far from any grid connection.

Solar power systems' integrate well with other energy sources and storage solutions, and their adaptability extends beyond location. Users can scale the systems for different needs:

>> **A few solar panels** can power a family's cabin.

>> **A commercial array** can offset a company's traditional energy use and save the company money.

>> **A utility-scale project** can produce hundreds of megawatts for the utility's customers' use.

This flexibility and scalability means that solar can support advanced applications such as powering data centers or desalination plants as well as low-tech and vital uses like pumping water for irrigation. In humanitarian contexts, portable solar kits can restore electricity in disaster zones within hours.

Gaining independence from fossil fuels sources

By generating your own electricity from solar energy, you reduce exposure to volatile oil and gas markets, supply chain disruptions, and politically driven price swings. For countries, this energy independence strengthens national security and reduces vulnerability to external pressures. Individuals and businesses get peace of mind and know that their power supply is local, stable, and mostly under their control.

In regions prone to natural disasters and grid instability, solar power paired with battery storage can keep critical systems running when the broader power grid goes down. This level of self-reliance is invaluable during prolonged outages, helping households, hospitals, and businesses avoid costly disruptions. And consider these advantages:

>> **On a macro scale,** shifting away from fossil fuels means fewer economic shocks from sudden fuel price hikes, fewer risks tied to foreign supply chains, and more control over national energy strategies.

>> **On a micro scale,** the use of solar power means not worrying about whether the lights will stay on when fuel prices spike or geopolitical tensions rise.

Small to Supergiant: Solar Does It All

Solar energy's versatility is one of its greatest strengths. Unlike other energy sources that may be locked into a particular scale — such as massive coal plants that must be built at utility size or small diesel generators that don't scale up efficiently — solar systems can fit just about any need. Solar systems are relevant to applications for homeowners, businesses, utilities, and governments alike, which ensures that they can play a role in nearly every aspect of the global energy transition. Table 1-2 provides a quick guide to how the solar power industry generally classifies project sizes. The boundaries aren't fixed, but they help distinguish the scale, economics, and applications of solar.

Residential projects

For households, residential solar has become a symbol of self-reliance, savings, and modern-living. A typical home system might be 3–15 kW, which is enough to cover a significant portion of a family's electricity use. The significance is not just about reducing bills, it's also about gaining a sense of control over energy in a world in which electricity prices often rise year after year, recently outpacing inflation. Figure 1-2 shows how the price of electricity has gone up since 1999 based on source data from the U.S. Energy Information Administration.

The adoption journey usually starts with a homeowner looking at the amount of their increasing utility bill and wondering whether solar power use can help. With available tools (such as the National Laboratory of the Rockies PVWatts or other online calculators), they can get an estimate of how many panels will fit on their roof, what the system will cost, and how long it will take to "pay back" through utility bill savings. Thanks to financing innovations like solar loans, leases, and power purchase agreements (PPAs), upfront costs are no longer the barrier to going solar that they once were. Many families install solar power systems with little or no money down and start saving immediately.

Once installed, residential solar power often brings an emotional payoff alongside the financial one. Homeowners can open an app on their phone and watch their system generate power in real time. It's empowering to see the Sun powering the refrigerator, the lights, or even an electric vehicle charging in the driveway. This sense of independence resonates across political lines; some see it as a green life-style choice, others as a way to cut ties with utility companies.

TABLE 1-2   ## Typical Solar Project Sizes and Categories

Category	Typical Capacity Range	Common Settings	Powering
Residential	3–15 kW	Homes, apartments, cabins	One household or building
Commercial	15 kW–5 MW	Warehouses, schools, offices	A single business or campus
Utility-Scale	5–500+ MW	Large solar farms (desert or farmland)	Entire towns or regions

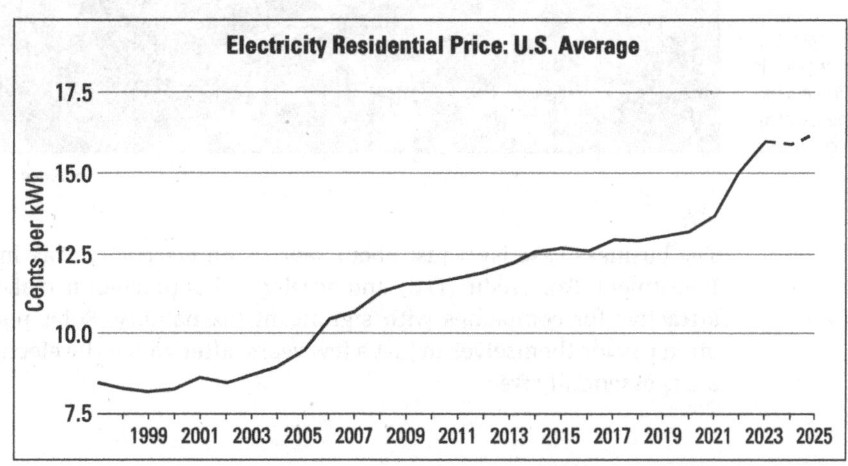

FIGURE 1-2: Increase in the residential electricity prices in the United States.

U.S. Energy Information Administration/Public domain

Another underappreciated benefit is home value. Studies have shown that homes with solar panels often sell faster and for more money than comparable homes without them. Buyers see the appeal of inheriting lower energy bills and a modern, sustainable upgrade. With home batteries becoming more common, residential systems can now keep the light on during outages, a feature that's increasingly important in areas prone to storm, wildfires, or grid failures.

Commercial projects

Businesses of all sizes are finding solar power systems to be a smart investment. Figure 1-3 shows an office building with solar panels on the roof. A mid-sized company might install a 250-kilowatt system on its warehouse roof and cut energy bills by 20 to 40 percent. Large corporations with sprawling campuses or retail footprints can save millions annually by deploying solar across multiple sites. Target, Walmart, and Apple are among the top corporate adopters, covering rooftops and parking lots with solar to cut costs and demonstrate going-green leadership.

FIGURE 1-3: Solar panels sprawling across the rooftop of a business.

azkiyanaila taleta/Adobe Stock Photos

The business case isn't just about savings on electricity. Tax incentives like the Investment Tax Credit (ITC) and accelerated depreciation make solar especially attractive for companies with significant tax liability. Solar power systems can often pay for themselves in just a few years, after which the electricity they generate is essentially free.

Some businesses go even further by combining solar with storage and on-site microgrids. For companies that can't afford downtime — like data centers, hospitals, or manufacturers — solar power provides resilience in addition to savings. These systems can keep critical operations running during outages, offering a competitive advantage in industries in which reliability is paramount.

Businesses benefit from both financial savings and improved reputation, which means that using solar power systems serves as both an economic and branding investment.

Utility-scale projects

At the largest scale, utility solar projects can generate hundreds of megawatts, enough to power tens of thousands of homes. These massive installations, sometimes called solar farms, are typically built on large tracts of land and feed directly into the grid. You may have seen one of these solar farms outside an airplane window! They are the backbone of the solar revolution, delivering clean energy at a scale that rivals or surpasses traditional fossil-fuels plants.

Utility-scale solar has driven the cost down through economies of scale. The bigger the project, the lower the per-watt cost of construction and operation. This is one reason that solar power has become the cheapest form of new electricity in many markets (see the section "Embracing solar as the most affordable form of energy" earlier in the chapter). For utilities, building large solar farms is often the fastest and most affordable way to add capacity to meet rising demand.

These huge projects, however, aren't without challenges. Land use is a major consideration: Should farmland be converted to solar fields? What about desert ecosystems or wildlife habitats? Developers are addressing these concerns through dual-use strategies and mitigation efforts discussed in the section "Land use and community concerns" later in this chapter. The strategies can turn land into a multi-use asset, supporting food production and clean energy at the same time.

Acquiring the necessary permits and interconnecting with existing power systems are also hurdles. Getting approval to build and connect a massive solar farm to the grid can take years and millions of dollars. But when successful, the payoff is huge: long-term, low-cost electricity for entire regions, along with job creation, tax revenue, and regional economic growth.

Energy storage is increasingly paired with utility-scale solar power systems to deliver power even when the Sun isn't shining. By adding large battery systems, utilities can smooth out fluctuations, provide power during peak evening demand, and reduce reliance on fossil fuel *peaker plants* (power plants that only run during periods of highest demand). This combination is a cornerstone of a modern, resilient grid.

Acknowledging the Dents in the Crown

Solar power may be wearing the crown in today's energy transition, but no king is without flaws. As powerful and transformative as solar energy is, it comes with real-world limitations that deserve attention. Understanding these challenges doesn't diminish solar's value, it equips you with a more balanced perspective and helps set realistic expectations for how and where solar power systems work best. Solar is an extraordinary tool, but like any technology, it works within certain boundaries and requires complementary solutions.

Regarding reliability and timing

One of the most common critiques of solar is that the Sun doesn't always shine. Solar power is intermittent: It produces electricity during daylight hours, peaks around midday, and drops off to zero at night. This mismatch between production and demand is sometimes called the *duck curve* (see Figure 1-4), in which electricity demand remains high in the evening just as solar output declines. Without solutions, this pattern creates stress on the grid and forces utilities to rely on backup power sources, often fossil fuels.

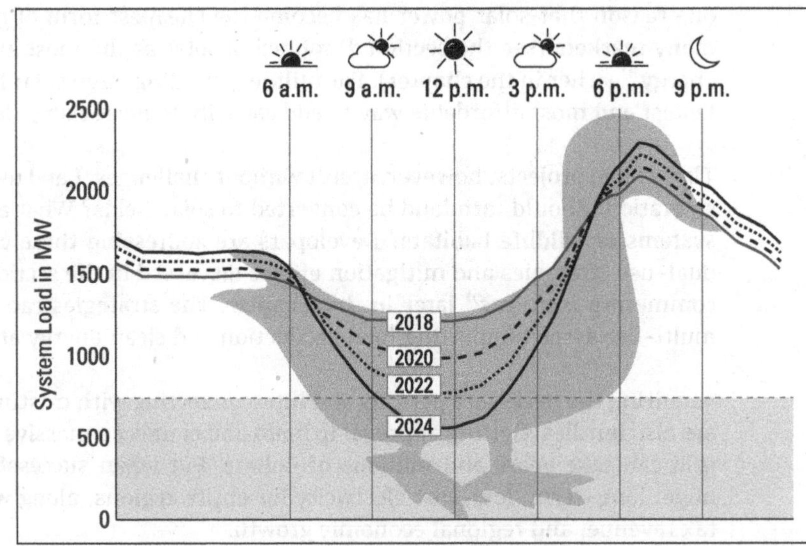

FIGURE 1-4: The duck curve demonstrates the imbalance between solar power production and energy requirements.

Battery storage is the most promising fix for the duck curve imbalance. By storing excess daytime energy and discharging it at night, batteries smooth out the curve and make solar power more reliable. Still, large-scale storage is expensive and technically challenging. A solar farm paired with batteries might provide four to six hours of evening coverage, but building enough storage for several cloudy days is another matter entirely. Other solutions may include smarter grids that shift demand to daylight hours (like running industrial processes during the day) and pairing solar with complementary energy sources like wind or hydropower and space-based solar power.

To see solutions to the duck curve in action, consider California, which has one of the highest levels of solar penetration in the world. The state's grid operators must carefully balance solar generation with evening demand surges, sometimes curtailing (shutting off) solar plants when the grid can't absorb the extra power at midday. This highlights how success at scale brings new challenges, making reliability a system-wide issue rather than a technical flaw of solar panels themselves.

Solar energy systems don't need to provide 24/7 power on their own, but can work great as part of a diversified energy mix.

REMEMBER

Cutting the red tape (or not)

Another dent in solar power's crown is bureaucracy. Building solar projects — whether rooftop systems or massive utility farms — requires navigating layers of permits, approvals, and interconnection agreements. In some places, homeowners face months of delays just to get approval to install panels. For developers of larger projects, the process can stretch into years, with environmental reviews, zoning hearings, and utility negotiations all adding complexity.

These hurdles slow down the pace of solar power adoption. A project that is technically and financially ready can stall simply because of paperwork. This isn't unique to solar projects, but it feels particularly frustrating when the technology is available and costs are low. In the United States, the permitting process for a rooftop system can sometimes take longer than the actual installation, which often takes only a day or two. For utility-scale projects, local opposition can derail plans entirely, even when developers are meeting legal requirements.

The good news is that many governments are working to streamline this process. Programs like "instant permitting" for residential solar are being piloted in some states, allowing homeowners to get approval in hours instead of weeks. On the utility side, some regions are creating clearer interconnection rules and timelines to prevent projects from languishing in regulatory limbo. These reforms show that while red tape is real, it's not insurmountable.

Land use and community concerns

While solar projects bring clean energy, jobs, and investment, they also raise questions about land use. Utility-scale solar often requires thousands of acres, and communities sometimes resist seeing farmland or open landscapes converted to rows of panels. In some cases, residents worry about property values, visual impacts, or effects on local wildlife. These concerns can lead to protests or lawsuits that can delay or even cancel projects.

Developers are responding with creative approaches to show that solar doesn't have to be an either/or decision between energy and other land uses:

>> **Agrivoltaics** combines farming with solar and allows crops or livestock to coexist with panels, turning potential conflict into collaboration.

>> **Pollinator-friendly solar farms** enable the growing of native plants under and around arrays to improve biodiversity while generating energy.

>> **Additional nearby land purchase** to hold the land for environmental or non-development use.

Don't let the challenges overshadow solar power's potential. Every energy source has trade-offs: Coal plants need constant fuel deliveries; nuclear plants face safety concerns; and gas pipelines can fail. Solar power's challenges are real but solvable.

Chapter **2**

Understanding Your Energy Use

Before you can figure out how solar fits into your life, you need to understand your baseline: your electrical energy use. This chapter equips you with the knowledge to interpret utility bills, identify where your electricity goes, and see how those patterns link to solar power opportunities. It's not just about numbers on a page, it's about uncovering insights into your daily routines and choices that affect how you use electrical energy.

Many people never look beyond the amount-due line on their bill, but hidden in those statements are clues about when you use the most electricity, which appliances drive your costs, and how your habits affect the amount-due line. By digging into these details, you begin to see where solar power can have the greatest impact and fit seamlessly into your lifestyle.

Understanding your energy use also connects you to the bigger picture. Every home is part of a larger energy system, and your patterns interact with those of your neighbors, your community, and even the wider electrical grid. When you recognize this linkage, you can discover not only how to save money but also how your choices support effective energy consumption. This chapter offers you the confidence to read energy-bill data like a pro and the perspective to see how solar power fits into the bigger story of a shared energy future.

Dissecting Your Energy Bills

Most people treat their electricity bill like a traffic ticket: They glance at the total, groan, and pay it. But hidden in that envelope or email are layers of information that can tell you a lot about your home, your habits, and your opportunities to save on energy costs. Reading your electricity bill is like reading a financial statement — at first the attempt feels like deciphering clues to a spy thriller, but once you crack the code, the story becomes obvious.

Understanding the content of your electricity bill is so important in the solar journey; this understanding gives you a baseline. Every calculation about how much solar power can save you, how big a system you might need, or how quickly you can recoup your investment starts with recognizing what you're paying now. If you skip this analysis, you're flying blind in the quest for a solar-power solution.

Before I get too deep into the electricity-bill weeds, I want to talk about two words that appear everywhere on a bill and in this book: power and energy. They sound similar but mean different things in this context:

>> **Power,** measured in kilowatts (kW) or watts (W) or even megawatts (MW), is like the speedometer in your car. It shows how fast you're using electricity at a given moment. That is, the W on the meter of your house shows how much power you are using in that moment.

>> **Energy,** measured in kilowatt-hours (kWh), is like the car's odometer. It tells you the total distance you've traveled, or in this case, the total amount of electricity you've consumed over time. That is, the kWh on your bill measures your energy use over time.

Your electricity bill usually charges you for energy (kWh), but sometimes it also sneaks in costs related to power (kW), such as demand charges based on your single biggest usage spike. Knowing this distinction makes the accounting of your energy use much easier to interpret.

Electricity bills can seem confusing because they reflect two levels of electrical costs:

>> **The cost of the energy you consumed,** which means the kWh flowing into your home. If you think of your energy consumption as dining out, this is the cost of the food you ate.

>> **The costs of the broader grid that delivers your energy.** That's why you see separate line items for energy, delivery and fixed fees. To continue the dining-out metaphor, the delivery fee is like a service charge to keep the restaurant open, and the fixed fee is like a cover charge at the door.

When you figure out how to separate these costs, the amounts you consider for calculating potential savings from solar power become much clearer.

To put the two levels of cost into perspective, imagine two neighbors with the same total cost on their bills. One uses lots of power at night when rates are high; the other uses most of their power during sunny afternoons when rates are lower. The total amounts may be identical, but the line items on their bills tell very different stories. When you figure out how to read those stories, you see why adding a solar power system impacts households differently and why your unique usage patterns matter so much.

Checking out a sample electricity bill

Even after you've looked at a dozen bills, the format can still feel overwhelming. Utilities companies pack a lot of line items into one or two pages, and each utility can present items differently. But underneath the design quirks, almost every bill contains the same building blocks. I walk you through a sample bill — line by line — so you know exactly what you're looking at. You see these general items:

>> **Service address and billing period:** This tells you which home the bill applies to and date range covered. Always sanity-check these dates; a "monthly" bill might actually cover 28 days one cycle and 33 days the next. The small change can affect comparisons.

>> **Meter reads and usage (kWh):** Bills show a starting meter reading and an ending reading and then subtract to calculate how many kilowatt-hours you used. If you see "estimated" instead of "actual," it means the utility didn't read your meter that month and made a guess. Treat those numbers with caution.

>> **Energy charges:** This is the core of your bill, the number of kilowatt-hours multiplied by the rate for that period. Some utilities keep it simple with a flat rate, but others break it into tiers (higher use costs more per kWh) or time-of-use buckets (off-peak, mid-peak, and peak hours each have different prices). See the next section for a description of various rate plans that affect the price you pay for energy.

>> **Delivery and transmission charges:** These charges are the costs of getting electricity from power plants to your home through substations and wires. They can be a flat fee, a per-kWh charge, or a mix of both. Delivery charges are often the part of the bill people forget about when comparing rates.

>> **Fixed customer or service charge:** Even if you used zero electricity, you'd still owe this fee. It helps cover the cost of keeping the grid connection alive, maintaining meters, and providing customer service.

>> **Taxes, riders, and credits:** States, counties, and utilities often add line items for things like renewable portfolio standards, energy efficiency programs, or infrastructure upgrades. On the flip side, you might also see credits here for participating in community solar, low-income assistance programs, or EV rebates.

>> **Demand charges (if present):** More common for commercial accounts, some residential customers are starting to see these charges, too. Demand charges are based on your highest 15-minute average kW draw during the billing cycle. Even if your total energy use is moderate, one big spike, like running the oven, dryer, and AC all at once, can raise this fee.

>> **Total due and due date:** Finally, the number most people skip to. Now that you know the items that make up your total amount due, it's no longer a mystery number.

Imagine two households and the differences that line items can make in their electricity bills and their potential for savings from adding solar power:

>> **A suburban family sees a $200 bill.** When they break it down, they realize $70 of it (35 percent) is fixed (delivery charges they can't avoid). The remaining $130 is tied to kWh usage, which adding a solar power system can help reduce.

>> **A small-apartment dweller sees a $90 bill.** Half of that amount is fixed fees, meaning even with added solar power or extreme conservation, the lowest their bill could ever go is around $45.

Collecting the energy usage data

Before you can make sense of your electricity costs, you need a clear and consistent record of your usage. Looking at just one or two bills won't cut it. For energy planning, you want to evaluate at least 12 months of bills. That gives you a full year of seasonal patterns: summer air-conditioning spikes, winter heating loads, and the transitional shoulder months of spring and fall in between.

If your utility provides an online portal, don't stop at grabbing PDFs of your bills. Many utilities allow you to download interval data, which can include detailed records showing your usage every 15, 30, or 60 minutes. This type of data is gold for solar planning because it reveals not just how much you use, but when. A household that runs heavy loads at night will see very different savings than one that consumes most of its energy during the day.

Many utilities participate in the *Green Button initiative*, which standardizes data downloads. Look for a green button labeled Download My Data on your utility company's site.

Here are the data that you want to gather:

>> **Basic bill info:** billing period start and end dates, total kWh, total dollar amount, and rate plan which may be

- Flat rate, you pay the same price per kilowatt-hour (kWh) no matter when you use electricity.

- Time-of-use (TOU), you pay different prices depending on the time of day. Power costs less during off-peak hours (typically midday or overnight) and more during demand periods, like the morning and evening.

- Tiered rate, you pay one price for your first block of usage and a higher price once you exceed that threshold.

>> **Charges and fees:** delivery charges, fixed fees, taxes, and credits

>> **Special charges:** demand charges (measured in kW)

>> **Interval data (if available):** timestamped kWh used, any notes for estimated reads, outage flags, and net exports, if you already have solar

Collecting all of this data might feel like overkill, but think of it as building your personal energy ledger. After you have it, you can slice and dice the data any way you want: by season, by appliance changes, or by the impact of new behaviors, such as EV charging. Table 2-1 offers a concise look at the data you need to capture and why it's important for solar planning.

TABLE 2-1   **Key Data to Capture**

Field	Why It Matters	Notes
Billing period start/end	Normalizes comparisons between months	Some months run longer than others
Total kWh	Monthly amount of energy used	Track seasonal highs and lows
Total amount due	Total monthly payment, the bottom line	Divide by kWh to get effective rate
Rate plan	Revels how prices change	Flat, TOU, or tiered
Delivery/fixed fees	Costs you can't offset with usage behavior	Important for minimum bill estimates
Demand charge (kW)	Shows spikiness, unevenness, of usage	Common in commercial, rare in residential
Taxes/credits	Capture discounts of solar/EV credits	Prevents overestimating true cost

Divvying up costs, month by month

After you gather a year (or more) of bills and extra data, the next activity is to make the numbers work for you. A stack of PDFs won't tell you much until you put the like data side by side. Think of this activity as turning your bills into a story, one that unfolds month by month.

Follow these steps:

1. **Create a simple spreadsheet or ledger, usually in a program such as Microsoft Excel or Google Sheets.**

2. **On the top row of the spreadsheet, add the column headings that identify the information you plan to include in your energy story.**

   For this accounting of your energy bills, you have these headings:

   - **Month**

   - **kWh** (for total energy used)

   - **Amount Due ($)**

   - **Effective $/kWh** (which you calculate in Step 3)

   - **Fixed & Delivery Costs**

   - **Percentage of Fixed Costs** (which you calculate in Step 3)

   - **Notes** (which include any comments about special events: outages, heat waves, houseguests, new appliances, and so on)

3. **Calculate the metrics that highlight patterns in your energy use for each monthly bill.**

   In this spreadsheet, you have columns identified for these calculations:

   - **Effective ($/kWh):** Figure this effective energy rate by dividing total **Amount Due** by total **kWh**. This calculation shows your true cost of electricity per kilowatt hour, not just the advertised rate.

   - **Percentage of Fixed Costs:** Figure the percentage of your bill that relates to **Fixed & Delivery Charges** versus the cost tied to your energy use by dividing the **Fixed & Delivery** amount by the **Amount Due ($)**.

4. **Fill in the spreadsheet with the data from your monthly bills and your calculated numbers from Step 3.**

   When you're finished, you'll have a document with information that looks something like Table 2-2.

Monthly Bill Ledger Example

Month	kWh	Amount ($)	Effective ($/kWh)	Fixed & Delivery Costs	Percentage of Fixed Costs	Notes
Jan	820	$156	$0.19	$35	22%	Cold, snow
Feb	800	$152	$0.19	$35	23%	Cold, snow
March	640	$128	$0.20	$35	27%	Mild weather
April	600	$120	$0.20	$35	29%	Mild weather
May	670	$141	$0.21	$35	25%	Mild weather
June	780	$187	$0.24	$40	21%	Warming up
July	900	$243	$0.27	$40	16%	Heat wave
August	850	$204	$0.24	$40	20%	Heat wave
September	740	$148	$0.20	$40	27%	Cooling down
October	680	$136	$0.20	$40	29%	Cooling down
November	750	$143	$0.19	$40	28%	Some cold
December	800	$144	$0.18	$40	28%	Cold, used heat pump

Gaining insight from your monthly data

With your ledger in place in spreadsheet form, you can achieve real insights when you make the data visual by creating a chart like the one shown in Figure 2-1. Use the features of your spreadsheet software to plot a bar chart for monthly kWh alongside a line for effective $/kWh. Immediately, you see whether your high bills are driven by using more energy or by paying higher rates in certain months.

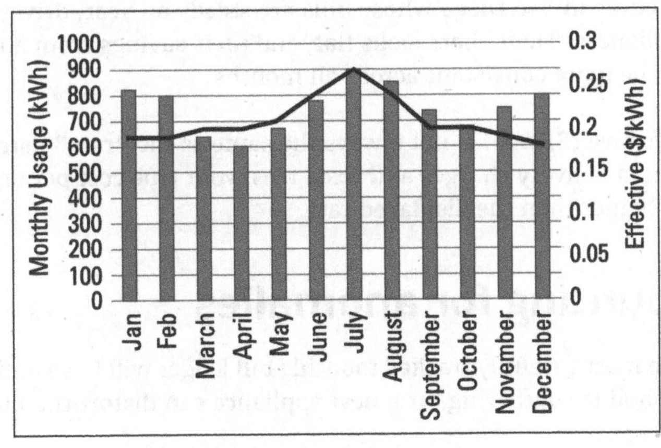

FIGURE 2-1: Chart showcasing the overlay between energy usage and total bill.

If you have pricing from a time-of-use (TOU) rate plan (as shown in Figure 2-2), this analysis is even more valuable. You might notice that your total kWh hasn't changed, but your bill's amount due has jumped because you're using more power during the peak (expensive) time window. If you have tiered pricing, you might see the opposite — your kWh stays steady, but creeping into a higher rate tier by using more energy pushes your costs up.

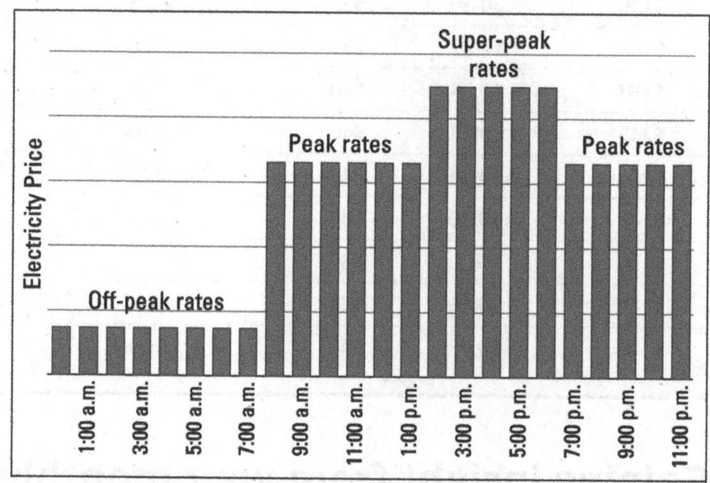

FIGURE 2-2: Sample time-of-use pricing chart.

Take a family of four in Phoenix and notice that their summer electricity bills skyrocket to $300–$400 because of air conditioning that runs relentlessly. By plotting their bills over the year, they may see that cooling (running the air conditioner) represents nearly half their annual electricity use. That knowledge helps them understand that adding a solar system won't necessarily trim costs evenly but instead, can dramatically cut their summer expenses. Contrast that with a single retiree in San Diego whose bills are steady all year, driven mostly by lights and appliances. Their chart looks flat, and their savings from an added solar system can be more consistent across all months.

Your effective ($/kWh) is not always the same as the "rate" listed on your bill. By blending in delivery charges and fixed fees, your true cost per unit of electricity is usually higher than the displayed rate.

REMEMBER

Accounting for anomalies

Even the most carefully tracked monthly bill ledger will have outliers. A freak heat wave, a holiday gathering, or a new appliance can distort the numbers and make your average usage look higher (or lower) than it really is. If you don't account for these anomalies, you risk oversizing or undersizing your solar system, which can mean spending more than you need to or falling short of your energy goals.

The trick for making a proper evaluation of your energy use is to spot unusual months, note why they're unusual, and adjust your analysis accordingly.

Common anomalies to watch for include

>> **Weather extremes:** Heat waves drive air-conditioning loads sky-high; cold snaps do the same with electric heating.

>> **Lifestyle changes:** Guests visiting for two weeks, kids home from college, or a new baby can all increase usage temporarily.

>> **Appliance events:** A new fridge, a broken pool pump timer, or an added EV can permanently shift your baseline.

>> **Billing quirks:** Sometimes utilities estimate usage instead of taking actual meter reads, or they might change your rate plan mid-cycle.

>> **Construction/renovation:** Power tools, portable heaters, or dehumidifiers can spike usage during a project.

Table 2-3 gives you an overview of common power-use anomalies and how you can make adjustments to the analysis of your energy use to discount them.

TABLE 2-3   ## Adjusting for Common Anomalies

Anomaly	Impact on Ledger	Simple Adjustment
Heat wave or cold snap	Higher kWh, possible higher $/kWh on TOU	Compare with weather data; don't treat as baseline
Guests staying	One-month bump in kWh	Note in comments, exclude from averages
New EV mid-year	Permanent jump in monthly kWh	Model EV usage separately going forward
Estimated read	Flat or odd usage pattern	Blend with following month, flag as estimate
Rate plan change	Sudden jump in $/kWh without more usage	Separate pre- and post-change averages and model post-change going forward

Suppose you have a household in Chicago. Most of the year, your electricity bills hover around $120. But one February, an Arctic blast doubles your heating demand and the bill spikes to $280. If you treated that month as normal, you could overestimate typical usage and potentially design a solar power system that's too large for your real energy needs.

Now compare that to a California household that installs a new EV charger in August. Their bill permanently rises by $60 a month. That's not an anomaly; it's a new baseline. Marking the change lets them adjust their calculations to include the energy needs of the car going forward.

Adding Up Typical Energy Usage at Home

After dissecting your bills, the next logical step is to understand what's driving those numbers. Your monthly kWh total is the sum of dozens of smaller pieces: heating, cooling, lighting, cooking, appliances, electronics, and more. Breaking usage into categories shows you which systems dominate and which ones are barely making a dent. This knowledge helps you size a solar power system realistically and identify where efficiency improvements can make the biggest difference.

Here are the usual suspects for where your energy goes:

>> **Heating and cooling your home environment:** Air conditioners, heat pumps, and electric furnaces often consume the lion's share of the energy your household uses, especially in climates with extreme summers or winters (or both).

>> **Water heating units:** Electric water heaters can quietly account for 15–20 percent of energy usage.

>> **Large appliances:** Refrigerators, washers, dryers, dishwashers, and ovens.

>> **Electronics and lighting:** TVs, computers, game consoles, LED or incandescent lighting.

>> **Miscellaneous plug loads:** This is a catch-all category for chargers, gadgets (such as cordless vacuums), and devices that sip electricity all day.

Table 2-4 gives you a quick look at an average breakdown of household energy usage.

Typical Breakdown of Residential Electricity Use (U.S. Average)

Category	Share of Annual Usage	Notes
Heating & Cooling	30–50%	Climate dependent
Water Heating	15–20%	Higher if household has multiple showers or electric water heater
Refrigeration & Cooking	10–15%	Includes fridge, freezer, oven, microwave
Laundry & Dishwashing	10–15%	Dryers are especially energy-intensive.
Lighting & Electronics	10–15%	TVs, computers, lights, game systems; (lower if using LEDs for lighting)
Miscellaneous Plug Loads	5–10%	Chargers and gadgets

National averages are useful, but no two homes have electricity bills that look the same. A family with a heated swimming pool will see very different numbers than a retired couple in a condo. Your bills tell you how much you used, but not what used it. To bridge the gap between knowing your total usage and where it was used, you can:

>> Review past appliance purchases of big-ticket items like HVAC, water heaters, and refrigerators. The documentation that comes with these items typically provides information about energy usage.

>> Use an energy monitor, such as a smart meter, to measure energy loads of specific items (like your electric water heater).

>> Compare your usage with utility-provided breakdowns.

Looking at averages across the United States

So, how much total electricity does the average American household use in a year? A major factor in determining this amount, of course, is the size of households. According to the U.S. Energy Information Administration (EIA): Average annual residential usage is approximately 10,500 kWh. Table 2-5 offers a look at the EIA's monthly energy-use averages for various sizes of homes.

TABLE 2-5

TABLE 2-5   Typical Household Usage Across Home Types

Home Type	Typical Monthly Range (kWh)	Notes
Small Apartment (1–2 ppl)	400–600	Minimal appliances, shared walls
Mid-Size Home (3–4 ppl)	800–1,200	Central AC or heat pump
Large Home (5+ ppl)	1,500–2,000	Pools, multiple HVAC systems

Geography also plays a big role in household electricity consumption, and the average usage of 10,500 kWh per year (875 kWh per month) disguises a huge monthly range:

>> Hot southern states like Texas, Florida, and Arizona often see much higher-than-average consumption, mostly driven by air conditioning. In some regions, summer electricity use doubles or triples compared to spring and fall.

>> Cold northern states like Minnesota or Maine may also see higher usage if homes rely on electric heating. Winter usage peaks can rival the summer surges seen in the South.

>> Milder coastal states like California or Oregon tend to have lower per-home usage, partly because of moderate climates and partly due to efficiency standards and smaller average home sizes.

What these ranges mean is that typical energy use is difficult to pinpoint. A household in Phoenix may easily use 1,500–2,000 kWh in July, while a similar home in San Diego might use half that. Table 2-6 shows monthly average electricity usage by U.S. region, and Figure 2-3 offers a visual comparison for the United States.

TABLE 2-6   Regional Ranges in Average Monthly Usage

Region	Average Monthly Range (kWh)	Key Driver
Southeast & Gulf Coast	1,200–1,800	Air conditioning
Midwest & Northeast	800–1,000	Electric heating in winter and air conditioning in the summer
West Coast	500–900	Mid climate, efficiency standards

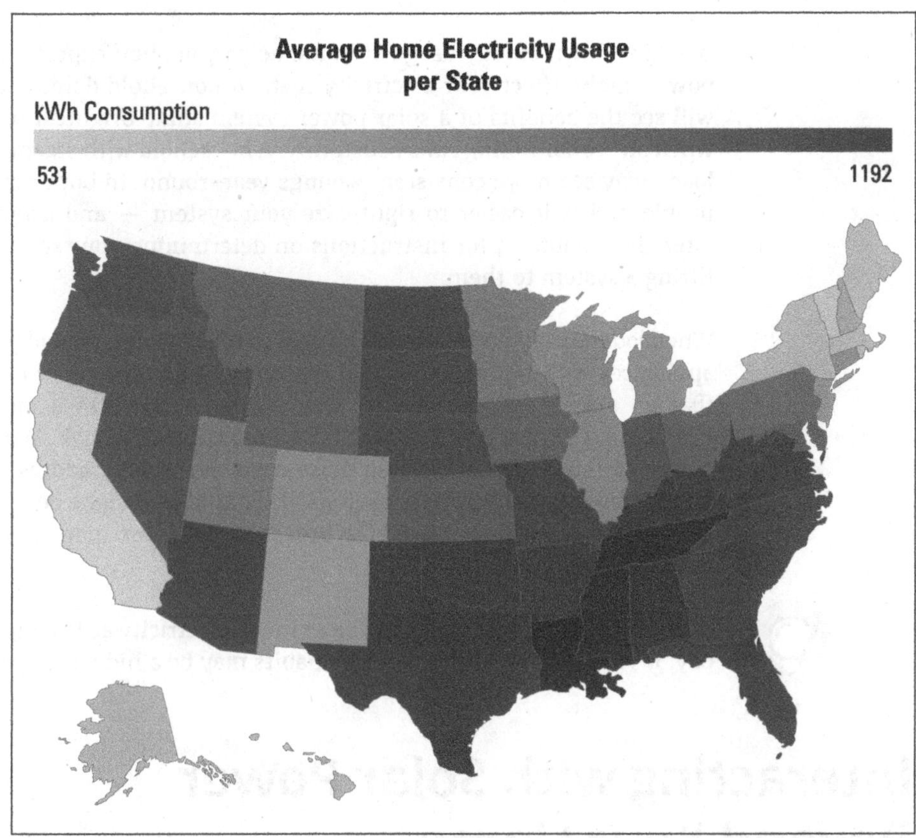

Created with Datawrapper/U.S. Department of Energy/Public domain

Relying on your specific energy use data

If you're curious about your state specifically, the EIA publishes detailed residential energy data. You can look up state-level averages for electricity use, prices, and even seasonal breakdowns. For the purposes of planning for adding a solar power system, though, it's often more useful to rely on your own bills than on broad averages because they reflect your home's unique story.

Your bills tell your truth. Averages are a starting point, but only your own 12–24 months of energy usage data will show how your household actually consumes electricity.

Breaking down energy usage at home helps you align expectations for how solar power might affect your electricity costs. A household dominated by HVAC loads will see the benefits of a solar power system shine brightest in summer months when air-conditioning runs constantly. A household with steady appliance-driven loads may see more consistent savings year-round. In both cases, knowing your profile makes it easier to right-size your system — and avoid disappointment later. See Chapter 4 for instructions on determining your solar power needs and fitting a system to them.

When looking at your bills with the goal to offset usage, consider swapping out old appliances with newer, more efficient ones. One homeowner in Texas discovered that an old garage refrigerator was adding nearly $25 a month to their bill. Replacing it with a new Energy Star model cut that in half. While adding a solar power system may have offset those costs eventually, addressing the appliance directly provided immediate savings. This kind of insight helps you see adding a solar power system as part of a broader toolkit for managing energy, not the only solution.

A single electric dryer load can use as much electricity as leaving your lights on all day. If your bills feel high, laundry habits may be a hidden culprit.

Interacting with Solar Power Beyond Your Home

When most people think about solar power, they picture rooftop panels offsetting their own electric bill. That's the most visible form of solar power systems, but it's only part of the story. Solar power is transforming how communities, businesses, and the entire electrical grid operate. Even if you never install panels on your roof, you can still benefit from, and interact with, solar power systems in ways that reach far beyond your front door.

Community solar projects pool resources from many households and businesses to build a shared solar farm. Participants subscribe to a portion of the output and receive credits on their utility bills, much like owning a slice of a bigger pie. It could be perfect for renters, apartment dwellers, or those with shaded roofs. You buy or subscribe to a share of the solar farm; then the electricity flows into the grid; and your bill reflects the credits (see Figure 2-4). This shared access is great because it expands the use of solar energy for people who can't install panels themselves.

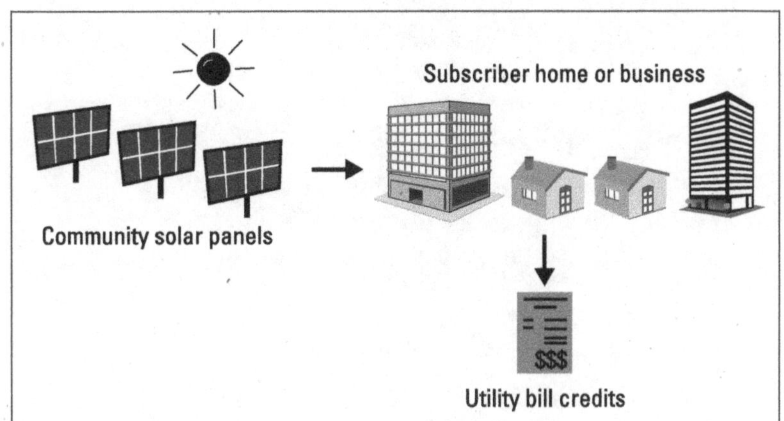

Every solar panel, whether on a home or a giant solar farm, connects — or ties into — the electrical grid. That means the energy you use may have been generated miles away on someone else's panels. Likewise, when your rooftop system produces more than you need, that excess can flow outward to your neighbors through the grid.

This *grid-tied setup* is what makes solar power so flexible and why utilities care about when and how much solar energy is produced. High penetration of solar systems can change the shape of demand curves, sometimes lowering midday demand so much that it reshapes how utilities manage power plants.

Chapter **3**

Pondering How the Sun Powers Our World

The Sun isn't just something that warms your skin on a summer afternoon; it's the ultimate powerhouse driving nearly every natural and human system on Earth. Plants grow because of sunlight; weather patterns form from solar heating; and every fossil fuel humans have ever burned started as *stored sunlight*. This chapter is all about tracing that energy back to its source and understanding how people can harness it directly.

In the chapter, you find out how solar energy travels across space in the form of radiation, how individual photons work their magic, and how the Sun's movement across the sky determines when and where solar power is most effective. You also get a clear picture of the two main flavors of solar power: *photovoltaic* (PV) panels that make electricity and thermal systems that capture heat. Finally, you follow the journey of a photon as it becomes usable power through a modern PV system, complete with the technologies and components that make it all possible.

These discoveries help you think about the Sun as more than a glowing ball in the sky: You can recognize the Sun as the most abundant energy source available to humanity.

Exploring the Powerhouse in the Sky

When you step outside on a clear day and feel the Sun warming your skin, you're experiencing just a fraction of the energy pouring toward Earth every second. The Sun is, quite literally, the ultimate powerhouse in the sky, and it's responsible for the forms of energy people use every day. Fossil fuels are simply ancient sunlight stored in plants and plankton that lived millions of years ago. Wind power takes advantage of the uneven heating (by the Sun) of Earth's surface. Even hydropower is driven by the Sun's activities related to evaporating water and driving the rain cycle. Without the Sun, Earth would be a dark, frozen rock drifting in space.

REMEMBER

The staggering fact is this: In just one hour, the Sun delivers more energy to Earth than all of humanity uses in an entire year. That means that even if humans capture only a sliver of it efficiently, solar energy could meet nearly all global energy needs. However, most of the solar energy pouring toward Earth is scattered, absorbed, or reflected before humans can capture it. Solar technologies are the way of directly tapping into this vast and continuous flow.

TIP

To wrap your head around the scale of the available energy, picture this: The energy hitting Earth's surface in a single day could power every home, car, factory, and data center on the planet for an entire year. It's a reminder that the problem isn't about supply, it's about finding better ways to capture and use what's freely raining down on us.

Acknowledging the origin and scope of available energy

The Sun itself is a gigantic nuclear reactor in the sky. Deep in its core, the Sun continually fuses hydrogen atoms into helium molecules through a process called *nuclear fusion*. This reaction releases unimaginable amounts of energy in the form of light and heat, which then travel outward through layers of plasma until they finally burst into space. It takes thousands of years for energy created in the Sun's core to reach its surface, but when it does, photons stream into space and reach Earth in just about eight minutes.

And the scale of the nuclear reaction is mind-blowing. The Sun fuses more than 600 million tons of hydrogen every second, producing enough energy to light up not only Earth but the entire solar system. And unlike the finite fossil fuels that humans dig out of the ground, this fusion process will likely keep going for another 5 billion years. For humanity, that means the Sun provides an essentially endless supply of energy waiting for people to devise smarter ways to catch and use it.

Solar radiation versus sunlight

People often use the terms *sunlight* and *solar radiation* interchangeably, but they're not the same thing. *Sunlight* is what people can see with their eyes — the familiar bright light that illuminates our days. *Solar radiation*, however, includes a much wider range of electromagnetic energy. Here is a breakdown of the complete solar radiation spectrum measured in wavelength units called nanometers (nm):

>> **Visible light (VIS, 400–700 nm):** What people and animals see as daylight in colors from violet to red, this band accounts for approximately 43 percent of solar energy.

>> **Ultraviolet (UV, 100–400 nm):** Invisible to human eyes, but this type of light is strong enough to cause sunburns and fade fabrics.

>> **Infrared (IR, 700 to ~1 nm):** Invisible to human eyes, but can be felt as heat, this band is a major component of the Sun's radiation, at about 49 percent.

>> **Other radiation:** Invisible to human eyes but emitted by the Sun, X-rays, gamma rays, microwaves, and radio waves are a tiny fraction of the radiation that reaches the Earth.

Together, the bands make up the solar radiation spectrum (see Figure 3-1), and solar panels are designed to capture certain slices of it. Most PV cells respond mainly to visible and near-infrared light, while thermal systems can use a wider range, including the ultraviolet, visible, and near-infrared portions.

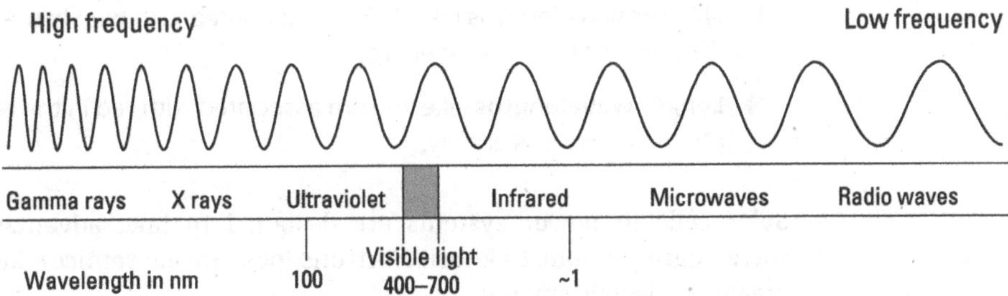

FIGURE 3-1: The solar spectrum.

Why does the information about solar radiation matter for you? Because where you live determines how much of the spectrum reaches your roof:

>> **A high-altitude city like Denver** gets more intense solar radiation (thinner atmosphere = less scattering of radiated energy).

>> **A coastal city with frequent fog, like San Francisco,** gets plenty of sunlight but less usable radiation on many days.

Clouds, pollution, and even the thickness of the atmosphere (known as *air mass*) affect how much solar radiation reaches your roof. That's why solar power potential in Denver looks different from solar power potential in Miami, even under the same Sun.

Finding out about photons

At the heart of solar power is the photon, the basic particle of light. A *photon* is a tiny packet of energy traveling from the Sun at the speed of light. You can think of it as the atom of light. Billions of photons strike every square centimeter of Earth's surface (only on the surface that's facing the Sun) each second. You can't see individual photons, but together they make up the sunlight streaming through your window.

Think of photons like raindrops. One drop is small, but billions together fill a bucket. In the same way, a single photon doesn't power much, but countless photons hitting a solar panel generate measurable electricity. Also picture a stadium wave at a sporting event. Each fan stands up and sits down, which is a small action on its own. But when 70,000 people do it in sequence, the effect is huge and visible across the entire stadium. Photons work the same way: individually tiny, but collectively powerful when captured by solar technology.

Each photon carries solar energy, and the amount depends on its wavelength. For example, for

>> **Shorter wavelengths** (like 400 nm associated with blue in the UV band), the photons have higher energy.

>> **Longer wavelengths** (like 760 nm associated with red in the IR band), the photons carry less energy.

Solar cells in power systems are designed to take advantage of the shorter wavelength photons to knock electrons loose inside semiconductor materials and create an electric current.

Imagine standing outside on a sunny day. The warmth you feel on your skin is mostly infrared radiation. The light that lets you see is visible radiation. And if you forget sunscreen, the sunburn later comes from ultraviolet radiation. All three arrive together as solar radiation (see the preceding section), but only some of the photons in the radiation bands are useful for electricity. That's why solar scientists and engineers work so hard to design materials that make the most of the spectrum.

Photons may be small, but their numbers are astronomical. Roughly 174,000 terawatts of solar energy reach Earth continuously, about 10,000 times global energy demand. Photons are the raw material, the building blocks, of solar energy. The challenge isn't whether they arrive — they never stop. But how effectively solar power systems can capture the photons is the challenge.

Plotting Your Sun Charts

One of the most important things about solar energy is that it isn't the same everywhere or at every time of year. The Sun rises, sets, and shifts its path across the sky depending on your location and the season. If you've ever noticed the Sun coming in through a different window in winter than in summer, you've already observed this. For solar energy planning, it's crucial to chart the Sun's movement and intensity — it's your *Sun chart*.

Charting out the basic path of the Sun

Earth is tilted on its axis by 23.44 degrees from the perpendicular to its orbital plane around the Sun, and that tilt is the reason that Earth experiences seasons. It also explains why the Sun doesn't take the same path across the sky every day or in every location on Earth. You find these differences:

>> **At the equator,** the Sun rises and sets almost directly overhead. Days are nearly equal length year-round.

>> **At mid-latitudes** (say, Kansas City or Rome), the Sun arcs high (closer to directly overhead) in summer and lower (closer to the horizon) in winter, which stretches or shrinks daylight hours. Figure 3-2 shows where the Sun's path falls on the solstices and equinoxes at mid-latitudes.

>> **At high latitudes** (think Alaska or Sweden), the swings are dramatic: You get the midnight Sun in June and barely any daylight in December.

Table 3-1 gives you an idea of how daylight hours change with season and location at various spots on Earth, each averaging 12 hours of sunlight on an annual basis.

Imagine two families: one living in Denver and the other living in Miami. In June, the Denver family gets more than 15 hours of daylight, while the family in Miami gets about 13. But in December, the daylight hours that the Denver family enjoys drops to only 9, while the family in Miami still enjoys 10.5. That's why Denver's summer solar power output dwarfs its winter numbers, while Miami's output is steadier.

Northern Hemisphere Timeframe	Sun and Earth Positions	Earth Axis Tilt and Result
Summer solstice (June)		Toward the Sun; direct light and longer days
Equinox (March and September)		Perpendicular to orbit; equal day and night
Winter solstice (December)		Away from the Sun; less direct light and shorter days

FIGURE 3-2: A diagram of where the Sun's path falls seasonally in mid-latitudes.

TABLE 3-1 ## Hours of Daylight Across Locations

City	Earth Location	Summer Solstice	Winter Solstice	Annual Average
Quito, Ecuador	Equator	~12 hours	~12 hours	12 hours
Denver, CO	Mid-latitude	~15 hours	~9 hours	12 hours
Stockholm Sweden	High-latitude	~18 hours	~6 hours	12 hours

Now imagine moving even farther north. In Stockholm, the summer solstice stretches to 18 hours of daylight, while the winter solstice squeezes it down to just 6 hours. Solar power systems can still work there, but their output swings wildly through the seasons.

If you want to check your own solar path, free apps such as SunCalc or SkySafari overlay the Sun's arc on a map that includes your home's location. You can liter-ally stand in your yard looking at the app and see whether your tree will block the Sun at 3 p.m. in October. That is, you can get an idea of when the sunlight will hit your roof (if you're thinking of adding solar panels) or your solar panels (if you already have them).

Sun charts aren't just about the Sun; they're also about what gets in the way. Trees, chimneys, taller buildings, and even hills can block sunlight at certain times. An obstacle that casts no shadow in June may shade panels for hours in December, when the Sun is lower. Check out Figure 3-3 to see how power produc-tion varies with outside influences.

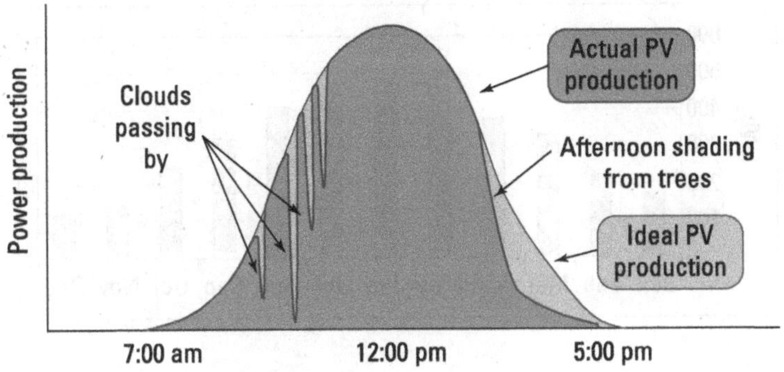

FIGURE 3-3: Production of solar power from photovoltaic cells showing the impact of shading and clouds.

Noting sunlight intensity

Even when the Sun is visible, its intensity changes. Scientists measure intensity as irradiance, in watts per square meter (W/m^2). On a clear day at noon, irradiance may hit 1,000 W/m^2.

A few considerations affect sunlight's intensity under various conditions:

>> **Solar noon:** Sunlight is strongest of the day and shadows are shortest.

>> **Morning or evening hours:** Sunlight is weaker because it passes through more atmosphere, which scatters photons.

>> **Clouds:** Thin clouds reduce irradiance a little; thick storm clouds can slash it dramatically.

>> **Air mass (AM):** This measure of how much atmosphere sunlight passes through varies by time of day. At noon, AM ≈ 1. At the horizon, AM can be 4–5, meaning sunlight is much weaker.

Think of sunlight supplying energy to your solar panels like using a hose to fill a bucket with water. At noon on a clear summer day in Phoenix, the hose nozzle is wide open, and the bucket fills quickly. But in Boston in December, the hose nozzle is barely trickling. Your solar power panels (the *bucket*) eventually fill any time of the year, but do so at a much slower rate in winter. In Figure 3-4 (for a home with four solar panels in mid-latitudes), notice that the height of the bars (signifying power production in kilowatt hours) is much lower in the winter months than it is in the summer months.

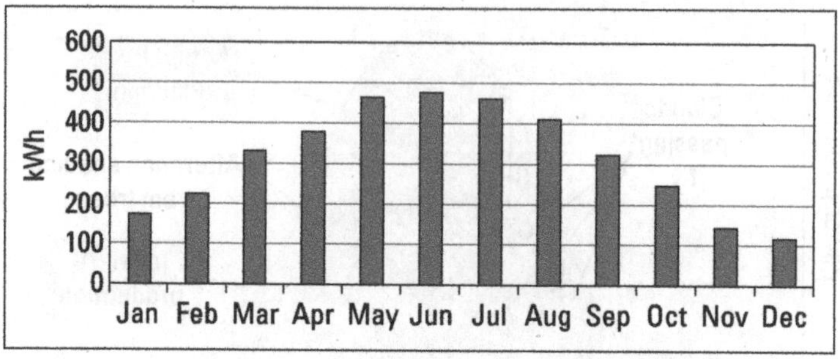

FIGURE 3-4:
Annual solar production and its change throughout the year.

Some surprising results can happen in different seasons and at various locations. For example,

>> Snow on panels reduces output, but snow on the ground can reflect sunlight upward, sometimes increasing production on clear days.

>> Summer sunlight is strong, but panels actually work less efficiently at high temperatures. A too-hot roof in Texas may reduce output even when the Sun is blazing.

>> Dust, smog, and fog in cities can cut irradiance significantly. That's why Los Angeles gets more sun hours than foggy San Francisco, even though they're at similar latitudes.

Plotting the Sun's path and measuring intensity explain why solar installers obsess over roof tilt, compass direction, and shading. These charts show not just how much sunlight your home gets, but when it gets it — and that's the key to predicting performance.

Solar power production isn't a flat number; it's a rhythm, changing with seasons, time of day, and even the location of a tree in your yard. Understanding the rhythms puts you ahead of the curve when it's time to design your own solar power system.

REMEMBER

Recognizing Photovoltaic versus Thermal Solar

When people say "solar," they often mean solar panels that make electricity. But that's only half the story. You can capture solar energy in two diverse ways: with photovoltaic (PV) panels (which produce electricity) and with solar thermal systems (which collect heat and pass the warmth to media such as air or water). These methods share the same energy source — sunlight — but the technologies, uses, and economics are distinct.

>> **Photovoltaic cells,** true to their name, work by turning light into electricity. PV panels are made of semiconductor materials, usually silicon. When photons strike the panel, they knock electrons loose from the semiconductor material and create an electric current. That electricity can power your home, charge a battery, or feed into the power grid. PV panels are modular and extend from rooftops to solar farms; these systems are probably what you see the most.

>> **Solar thermal systems,** on the other hand, don't make electricity. Instead, they capture the Sun's heat directly. A simple version of a thermal system is a black water tank on a roof, which is warmed by sunlight to provide hot water. Larger systems can use mirrors to concentrate sunlight onto a fluid, producing steam that spins a turbine.

You can easily confuse thermal solar systems with PV systems at a first glance, but to a trained eye, they look quite a different. PV panels are flat and often look blue or deep navy with metal grid lines that make up the cells. Thermal panels often look like glass covered boxes or tubes, sometimes even with water pipes visible. See Table 3-2 for a comparison of PV and solar thermal technology features.

TABLE 3-2   **PV versus Thermal Solar at a Glance**

Feature	PV (Photovoltaic)	Thermal
Output	Electricity/electrons	Heat of water, air, or fluid
Best For	Homes, businesses, utilities	Hot water, pools, industrial heat
Efficiency	15–25 percent of sunlight converts to electricity	30–80 percent of sunlight converts to heat)
Storage	Batteries	Insulated tanks

Imagine two neighboring families. One family installs PV panels on their roof, and the panels power their lights, refrigerator, and laptop computer. The other family installs a solar thermal system, which doesn't run their electronics but gives them nearly free hot showers and a warm swimming pool. Both families collect and use solar energy in completely different ways.

PV solar shines as a power source when you need electricity, and thermal solar shines when your heat demand is high and producing heat is your end goal.

PV is the dominant technology today. Globally, in 2026, PV capacity is more than 2,200 GW, while solar thermal power plants add up to only about 7 GW. But thermal systems can still make sense for specific applications.

PV and thermal aren't competitors so much as complementary tools. PV dominates the market because electricity is so versatile as a power source. But thermal has its niche, especially for heating water cheaply and efficiently. Knowing the difference helps you set realistic expectations, whether you're thinking about cutting your utility bill or just heating your pool.

Going from Photon to Power: How Solar Panels Work

Earlier in the chapter, you can find out what photons are and how the Sun delivers energy. But how does that invisible stream of light actually become the electricity that powers your phone charger, your air conditioner, or your neighborhood? In this section, you can trace the journey of a photon from striking a panel to flowing as usable power.

The journey from photon to power isn't magic; it involves physics, engineering, and smart design working together. Different panel types capture light in different ways. The full system (which includes panels, inverters, wiring, support structures, and batteries) transforms that captured light into usable energy. And after installation, system performance depends on practical factors such as roof angle, temperature, and shading.

The bottom line is that solar panels allow us to grab a stream of photons that left the Sun eight minutes ago and immediately put them to work in our homes and businesses. When you understand how that journey happens, you can make sense of why solar systems are designed the way they are, why production varies, and what to expect from your own panels over decades of use.

Types of solar technologies

When you read about the efficiency of a solar power system, you find out about the percentage of sunlight hitting the system that's actually converted to the desired energy output. For PV panels, that output is the production of electricity. Figure 3-5 shows what the various panels of the few primary types of solar panel technologies look like:

>> **Monocrystalline silicon (mono):** Made from a single, continuous crystal, mono offers sleek black panels and high efficiency (20–24 percent), but it's more expensive than other types of panels.

>> **Polycrystalline silicon (poly):** Made from multiple silicon fragments melted together, poly panels have a blue speckled look and are slightly less efficient (15–20 percent), but they're cheaper than mono panels.

>> **Thin-film:** Ultra-thin layers of photovoltaic material (like *cadmium telluride*, a compound semiconductor known for excellent light absorption), thin-film solar is flexible and lightweight, but lower in efficiency (10–13 percent of sunlight converted). This technology is great for special uses such as solar shingles or curved surfaces and is generally as low in cost as a traditional panel.

>> **Other emerging technologies:** Perovskites (compounds with a specific crystal structure) and tandem cells (which stack two or more solar cells that capture varying ranges of the solar spectrum) promise higher efficiency and lower costs but are still on the move from labs to markets.

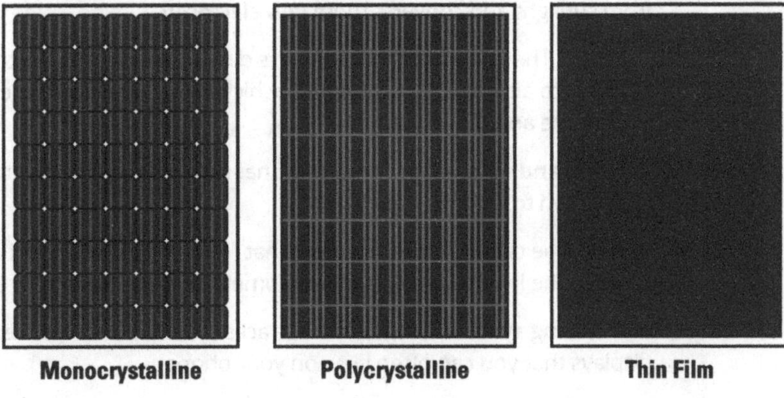

FIGURE 3-5:
Solar panel technologies.

Table 3-3 gives a quick comparison of solar panel technology.

TABLE 3-3   **Comparing Solar Panel Types**

Technology	Typical Efficiency	Cost	Look	Best For
Monocrystalline	20–24 percent	Higher	Navy blue, uniform cells	Maximum output
Polycrystalline	15–20 percent	Lower	Blue, speckled	Lower cost installs, larger areas
Thin-film	10–13 percent	Varies	Flexible	Unique applications, light structures
Emerging (perovskite/tandem)	25–30 percent+	TBD	Varies	Next-gen high-performance systems

Anatomy of a PV system

A solar panel doesn't work alone. It's part of a system that converts sunlight into electricity that you can actually use. The system includes these components:

>> **Solar modules (panels):** The visible part of the system that captures photons from sunlight and converts them into electricity

>> **Inverter:** The apparatus that converts direct current (DC) electricity from panels into alternating current (AC), which is the type of electric current that your home and power grid use

>> **Racking and mounting:** Structures that hold panels at the right tilt and orientation to catch the sunlight

>> **Wiring:** The cables and connectors that move electricity from the panels through the inverter and into your home's electrical system or the grid

>> **Monitoring system:** Software that tracks system performance, with related displays that you can often view on your phone

Think of the functioning of the system and its output like preparing food in a kitchen. The panels are the ingredients, the inverter is the stove that transforms the ingredients, the racking is the cabinets that hold the cookware and ingredients, and the wiring is the cookware that holds the ingredients that are cooking. Each part has a role in turning raw sunlight into something usable.

Looking at impacts on performance

Even the best system doesn't produce the same amount of electricity every day, and a big part of solar power's story is understanding what affects system performance. Here are some specifics:

>> **The angle and orientation of your roof make a huge difference. For roof-mounted systems,** in the Northern Hemisphere, south-facing roofs generally produce the most power because they get the longest exposure to the midday Sun. East- and west-facing roofs work, too, but they capture more sunlight in the morning or afternoon and less at noon, which lowers total output.

>> **Shading is a critical factor.** A single chimney shadow or a nearby tree branch can reduce output far beyond the small area shaded. That's because most panels are wired in series, like holiday lights — one shaded panel can drag down the whole string. Installers often use special tools to map shading throughout the year to avoid this problem.

>> **Temperature matters.** Counterintuitive as it sounds, panels don't like to be hot. Their efficiency drops as they heat up, so a rooftop baking at 100 degrees F may generate less power than the same panel on a cooler day.

>> **Dust, pollen, and snow can block sunlight,** though rain often takes care of routine cleaning.

>> **Over decades, panels gradually degrade,** usually by about half a percent per year. That means after 25 years, a panel might still be producing around 85 percent of what it did when it was brand new.

Table 3-4 offers a quick look at the impact of these various factors on system performance and efficiency.

TABLE 3-4 **Factors That Impact Solar Performance**

Factor	Typical Impact to Performance	Example Scenario
Orientation	20–30 percent	East-facing versus south-facing
Shading	5–50 percent	Tree branch shading part of a system
Heat	10–15 percent	100 degrees F rooftop in Phoenix
Dust/snow	5–20 percent	Heavy pollen, winter snow cover
Degradation	~0.5 percent annually	85 percent output after 25 years

2

Investing in Residential Solar: Powering Your Home

Chapter **4**

Deciding Whether Solar Power Is Right for Your Home

Figuring out whether solar power is right for your home isn't just about wanting clean energy or cutting your utility bill. It's about understanding how the pieces fit together, what size system your house can support, how much electricity it will realistically generate, and how those numbers translate into dollars saved each month. You need to develop a grounded framework for judging whether installing a solar power system on your roof makes sense for your home, financially, practically, and personally.

In this chapter, you explore the core benefits homeowners look for when they invest in solar power systems, from reducing long-term energy costs to gaining more independence from the utility company. Then you dive into the nuts and bolts of system sizing and production, where factors such as roof tilt, shading (from trees in your yard or neighboring buildings), and panel type make a measurable difference. And you see how to run quick calculations on system cost and savings, compare those results to your current electricity bill, and determine a likely payback period.

Recognizing the Benefits of Powering Your Home with PV Systems

The benefits of solar electricity for your home are numerous and compelling. Some are financial, some are about lifestyle, and others are simply about peace of mind. Although solar technologies have advanced and costs have dropped since the first edition of this book, the underlying reasons homeowners go solar have stayed remarkably consistent. These reasons include

>> **Lowering energy bills.** Every kilowatt-hour (kWh) of power your panels generate is one less kWh you need to buy from the utility company. Over the life of a system — typically 25 to 30 years — that savings can add up to tens of thousands of dollars in avoided costs. Solar also provides a hedge against future rate hikes. Utilities may adjust their prices over time, but your panels quietly generate power at no extra cost once they're installed.

>> **Gaining power reliability.** When paired with a battery, a solar power system can provide backup power during blackouts, keeping essentials such as lights, refrigerators, or medical devices running. Even without extra storage, knowing that your solar system can supply power for those household energy needs — directly from the Sun — gives many homeowners confidence and a sense of independence.

>> **Increasing property value.** Multiple studies show that homes with solar systems sell more quickly and at higher prices than comparable homes without them. Buyers increasingly see rooftop solar as an attractive upgrade, much like a renovated kitchen or a finished basement, except this upgrade pays for itself over time.

>> **Benefitting the environment.** Every kilowatt-hour of power produced by solar is one less generated from fossil fuels. While people weigh that factor differently depending on their values, the result is the same: Solar reduces emissions, lowers air pollution, and contributes to a more sustainable energy mix.

After you install a solar power system, you can start saving immediately. With every sunny day, your roof produces energy you'd otherwise be buying from your utility company.

Determining System Size

The most important factor in determining whether solar makes sense for your home is the size of the system you install. If the system is too small, you won't see much impact on lowering your utility bill. If the system is too large, you may spend more money to buy and install it than you'll ever get back. The sweet spot is finding a system that meets most of your needs without overshooting.

Start with your electricity bill and follow these steps:

1. **Add up 12 months of usage in kilowatt-hours (kWh) as outlined in Chapter 2, to get your energy annual demand.**

   A typical U.S. home uses somewhere between 8,000 and 12,000 kWh per year, although your demand could be higher or lower depending on climate, heating and cooling systems, and lifestyle.

2. **Translate the annual usage calculated into a required system size by dividing annual demand (in kWh) by the system's yearly capacity.**

   Across much of the United States, each kilowatt (kW) of solar capacity generates on average between 1,200 and 1,500 kWh per year, depending on location and conditions. So, if you use 10,000 kWh annually, make these calculations:

   $$10{,}000 \text{ kWh} \div 1{,}500 \text{ kWh} \approx 6.67 \text{ kW}$$

   and

   $$10{,}000 \text{ kWh} \div 1{,}200 \text{ kWh} \approx 8.33 \text{ kW}$$

   You can see that a system in the 7- to 9-kW range of solar power capacity will usually do the job.

Figuring to offset (not zero out) your electricity costs

Most homeowners naturally assume the goal is to hit 100 percent of their electricity use with solar. Why wouldn't you want to zero out your bill? The reality is that covering every last kilowatt-hour isn't always the smartest financial move.

In many states, utilities no longer credit you at the full retail rate for electricity you send back to the grid. That means if your system regularly produces more than your household consumes, the excess power may only earn you wholesale-level credits, a fraction of what you pay when you buy electricity. In other words, those extra panels might look good on paper but won't actually speed up your payback.

Many installers now recommend targeting somewhere between 70 and 90 percent of your annual usage. At that level, your solar covers nearly all of your daytime demand, trims down your monthly bills significantly, and avoids the trap of over-spending on panels that generate excess power that doesn't pay you back. If you add a battery, you can push your coverage higher because that stored energy reduces reliance on the grid in the evenings. Table 4-1 shows you rough estimates of the system-size you need based on your annual use and target electricity bill offset of 85 percent.

TABLE 4-1   **Rough System Size Estimates Based on Annual Use**

Annual Usage (kWh)	85 Percent of Usage for Offset	Average Annual Capacity (kWh)	Typical System Size (W)	Number of Panels (at 400 W each)
8,000	6,800	1,200–1,600	6,000–6,500	16
10,000	8,500	1,200–1,600	7,500–8,000	20
12,000	10,200	1,200–1,600	9,000–9,500	24

Of course, the system you should get isn't always the one you can get. Roof space, shading, and budget all place limits on what's possible. A small roof may only fit a 5-kW array unless you choose higher-efficiency panels. A heavily shaded lot may reduce production so much that even a bigger system underperforms. And sometimes budget is the final governor, even if your roof could hold 12 kW of panels, your wallet may be happier with 6 kW.

Planning for future needs

Another consideration for the size of the solar power system you buy is your future needs. For example, if you're planning to buy an electric vehicle (EV) or thinking of swapping out a gas furnace for an electric heat pump in the next couple of years, consider those applications when you determine system size. Those choices will increase your usage, and it's often cheaper to size up now than to add panels later.

System sizing is as much art as science. Installers rely on your past energy use, site conditions, and software tools to propose a number, but you get the final say on whether to go bigger, smaller, or right down the middle.

Comparing bids for your system

Solar panels naturally produce direct current (DC) electricity, but that's not what powers your home. In the United States (and most of the world), the grid and household appliances run on alternating current (AC). That means every solar system needs an inverter to act as a translator, converting the DC coming off the roof into the AC your house can actually use. See Chapter 3 for more information about the components of a solar power system.

Because of that conversion step, you'll see two different numbers in solar power proposals. The DC rating is the sum of the panels' nameplate capacity, basically what they could produce under perfect laboratory conditions. The AC rating is what's left after the inverter has done its work, and it usually comes out 10 to 20 percent lower than the DC capacity. Both numbers are accurate, but they tell different stories: one about raw potential, the other about usable output.

When comparing bids, don't be thrown off if one installer quotes a 7–kW system and another calls the same system 6–kW. They may just be using different conventions. For everyday purposes, the AC number is the more practical figure, because it represents the electricity that will actually flow into your outlets.

Predicting System Production

After you know how big your system should be, the next question is how much energy it will actually produce. On paper, it's tempting to think a 7-kilowatt system is just a 7-kilowatt system, no matter where you put it. In reality, production depends on a long list of details, everything from which panels and inverters you choose to how your roof is angled and whether there's a big oak tree nearby.

Solar system performance is influenced by both equipment choices and site conditions.

>> Equipment includes the type of panels you pick, the inverter that converts their electricity into usable power, and how the system is mounted.

>> Site conditions include your roof's tilt and direction, the amount of shading it gets throughout the day, and even how far the panels sit above the roof surface.

Each factor may sound small, but together they determine whether your panels deliver as expected or fall short.

When installers design a system, they use software to model all these variables and estimate annual production. For homeowners, understanding the basics of what goes into those models helps you spot whether a proposal makes sense and whether one installer's assumptions are more realistic than another's.

Considering panel efficiency

Not all solar panels are created equal, and the type you choose affects how much energy your system can realistically deliver. You get an overview of how different panel technologies work in Chapter 3. Here, the focus is on what those differences mean for your electricity production numbers.

High-efficiency monocrystalline panels produce more power per square foot. That means they're especially valuable if your roof space is limited, because you can squeeze more production into a smaller area. Standard-efficiency panels, often a bit cheaper, may require a larger footprint to hit the same production numbers.

If your roof has plenty of space and little shading, you may not need the most efficient panels on the market to meet your energy goals. But if your roof is small, partly shaded, or has awkward angles, investing in higher-efficiency models can be the difference between offsetting 70 percent of your usage and covering 90 percent.

In real-world terms, the difference between panel types may be a few hundred kilowatt-hours per year per kilowatt of installed capacity. That doesn't sound like much until you realize it can add up to thousands of kilowatt-hours, and thousands of dollars, over the system's life.

Don't assume "more efficient" is always "better." If you have ample roof space, mid-range panels may get you the same results for less money.

Looking into inverter types

Every solar panel on your roof produces direct current (DC), but your house runs on alternating current (AC). That's where the inverter comes in. The inverter is the brains of the system, converting DC into AC that your appliances and the grid can use.

In Chapter 3, you find out about the inverter as an essential component of a solar power system. Here, the key question is how their technology affects your system's production. Table 4-2 gives you a quick look at your inverter choices, including

>> **The traditional string inverter,** in which all the panels are wired in series, and the inverter converts their combined output into AC. The downside is that if one panel is shaded or underperforms, it drags down the whole string.

>> **Microinverters,** which involve putting a small inverter on each panel, solve the problem of an underperforming panel slowing things down. Every panel operates independently. If one panel is shaded by a chimney or tree branch, the rest still work at full strength. This makes microinverters especially helpful on complex roofs with multiple angles or shading issues.

>> **A power optimizer paired with a central inverter,** which gives you a middle-ground choice. Each panel gets its own optimizer, which conditions the DC before sending it to the inverter. Like microinverters, optimizers allow individual panel performance, but the system still relies on a single inverter for the final DC-to-AC conversion.

TABLE 4-2   **Comparing Inverter Options**

Inverter Type	How It Works	Where It Works Best
String inverter	One central inverter converts output from all panels in a series.	Unshaded roofs with a uniform panel layout
Microinverters	Small inverter on each panel converts DC to AC independently.	Roofs with shading and/or multiple angles or plans to expand system over time
Power optimizers	Device on each panel conditions DC before sending it to a central inverter.	Mixed-shade roofs, budget-conscious homeowners who want some panel independence

From a power production standpoint, the difference between these approaches can be significant. A roof with partial shading or panels split between east and west slopes may generate 10 to 20 percent more energy with microinverters or optimizers compared to a basic string inverter. On a simple, south-facing roof with no shading, though, the gains are smaller, and a string inverter may be the most cost-effective choice.

Evaluating the tilt angle

The Sun's path across the sky changes with the seasons and affects the intensity of sunlight hitting the Earth and, therefore, your house. (See Chapter 3 for more information on the effects of the Sun's path.) And understanding how and when the sunlight hits your roof pays off because the tilt angle of your panels determines how well they line up with the Sun's path, and therefore how much energy they'll produce.

The *tilt angle* is simply how steeply your panels are slanted relative to the ground. The rule of thumb is that the best angle for maximum annual production is close to your latitude, so if you live at 35 degrees north latitude, a 30–40 degree tilt is close to perfect. Figure 4-1 depicts sunlight hitting solar panels at two tilt angles.

But here's the good news: Solar is pretty forgiving. Panels tilted 20 degrees or 45 degrees may lose only a few percent of annual output compared to those at the "perfect" angle. That's why installers usually just follow the slope of your existing roof instead of building complicated racks to hit an exact angle. The production you lose isn't worth the extra cost or awkward roof profile.

Tilt can matter more if your utility pays different rates at different times of year. For example, a steeper tilt (closer to 40–45 degrees) can boost winter output, which may be valuable if your rates spike in colder months or you use a lot of electricity for heating. Shallower tilts (15–25 degrees) favor summer output, which helps in air-conditioning-heavy regions.

In real-world systems, the "perfect" angle is often just whatever keeps the panels flush with your roof. Ground-mounted arrays, though, give you more freedom to dial in tilt. That's why you'll often see panel arrays set on racks aimed just right for their location.

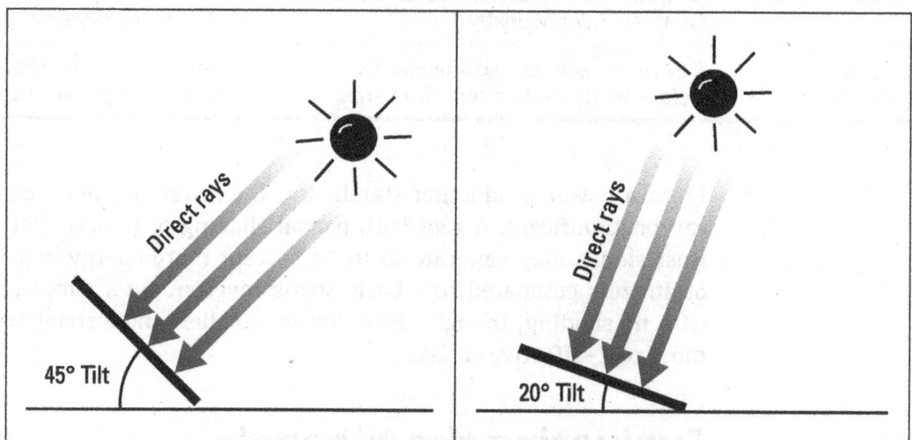

FIGURE 4-1: Example tilt angles.

Facing the azimuth question

If tilt is about how steep your panels sit relative to the ground (see the preceding section), azimuth is about which way they face. *Azimuth* is the compass direction your panels point, measured in degrees from north (see Figure 4-2). A south-facing azimuth (180 degrees in the Northern Hemisphere) usually delivers the highest annual production.

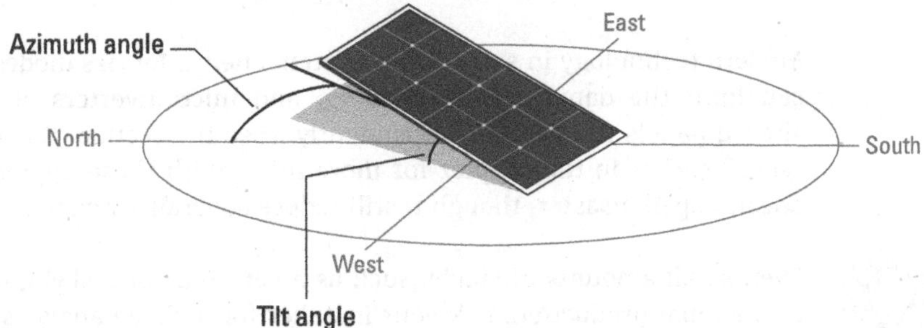

FIGURE 4-2:
Azimuth and tilt angles.

Azimuth indicates how your roof lines up with the arc of the Sun across the sky. A roof that faces due south (again, in the Northern Hemisphere) will catch the most direct rays of sunlight throughout the year, while east- or west-facing roofs will miss some of the peak sunlight hours. The difference isn't usually catastrophic: Solar installations on east or west roof orientations may produce 10 to 20 percent less energy than a south orientation, but the solar power output can still be worthwhile in many cases.

North-facing roofs, however, are a poor choice for installing solar power systems in the United States. The panels simply don't get enough direct sunlight to justify the cost. If your only available roof space faces north, you may need to consider a ground-mounted system or a creative design with multiple roof faces.

TECHNICAL STUFF

There's also a strategic twist to roof orientation. Some homeowners on time-of-use (TOU) rates actually benefit from a west-facing array because it produces more power in the late afternoon when electricity is most expensive. In those cases, a slightly lower total output can translate into higher bill savings.

Detecting shade levels

Tilt and azimuth are about geometry, but shading is about reality. Even if your roof has a perfect pitch and faces due south, shade can kill solar power production. A tree branch, chimney, or neighboring building that casts a shadow at the wrong time of day can drop output more than you'd expect.

Solar panels are built from strings of cells. When a cell is shaded, it acts like a bottleneck, reducing current flow through the whole string. In a traditional string inverter system, one shaded panel can pull down the performance of all the others in that circuit. That's why installers pay such close attention to shading during their site surveys. And sunlight intensity changes over the course of a day. Shading at the time when the Sun's intensity is the highest makes that effect even more pronounced. A single obstruction at 3 p.m. could mean you lose out on some of the day's most valuable hours of generation.

Modern technology in solar power systems helps. Bypass diodes built into panels can limit the damage from shading, and microinverters or optimizers allow shaded panels to operate independently. (See the section "Looking into inverter types" earlier in the chapter for more info.) With these upgrades, a little shade doesn't spell disaster, though it still reduces overall efficiency.

Even small amounts of shade, such as a vent pipe or a skylight shadow, can eat into annual production. Ask your installer for a shade analysis that models how obstacles affect output hour by hour and season by season.

Planning the distance from the roof

Solar panels work more efficiently (oddly enough) when they're kept cool. Heat reduces their ability to generate electricity, so installers mount them a few inches above your roof rather than flush against the surface. That air gap between the roof and the panels allows natural airflow to circulate and carry away heat build-up.

The difference in output production isn't dramatic (nothing like 50 percent), but it's measurable. A panel running hotter on a still summer day may put out 5 to 10 percent less than the same panel on a cooler, breezy afternoon. Over the course of a year, that efficiency loss adds up.

For most homeowners, this is a detail you don't need to fuss over; installers know the optimal clearance for your roof type and racking system. But that's why solar panels don't sit flat against shingles — that little space makes them last longer and work better.

A little breathing room keeps your panels cooler and more productive, and your roof stays healthier, too.

Distinguishing DC and AC ratings

Every solar power system comes with two measurements of capacity: DC and AC (see the section "Comparing bids for your system" earlier in the chapter). The DC rating is the sum of the nameplate capacity of all your panels, which is the maximum power they could generate under perfect lab conditions. The AC rating is what's left after the inverter does its job, converting DC into electricity your home and the grid can actually use.

For example, an array rated at 4.75 kW DC may deliver only about 3.9 kW AC after you factor in the conversion losses. That doesn't mean anything is "wrong" with the system; it's simply how physics and electronics work. Heat, wiring, and inverter efficiency all shave a little off the raw DC panel capacity numbers.

From a production perspective, this matters when you're comparing system quotes. Some installers highlight DC capacity, which makes the system look bigger. Others lead with AC capacity, which better reflects the power that you'll actually receive. Neither is misleading, but you want to be sure you're comparing apples to apples.

Over time, AC is the more practical number to track, because it aligns with your utility bill and household consumption. Still, knowing both figures gives you a fuller picture of how your system performs under real-world conditions.

Figuring expected energy output

When you consider all the factors of your solar power system — the panel type you chose, the inverter design, the tilt and azimuth of your roof, and how much shading exists — you arrive at the real number that matters: how much electricity your system will actually generate. Each of those factors tweaks the total. A south-facing roof at the right tilt angle in Phoenix will deliver more kWh per kW capacity than a shaded east-facing roof in New Jersey, even if both have the same "10 kW" system size on paper.

Modeling your output

Installers use modeling software (such as PVWatts, HelioScope, PVSyst, or Aurora) to make predictions about the expected energy output, but you can understand the basics being measured without running the software yourself. Across the United States, one kilowatt of solar capacity can produce between 1,100 and 1,800 kilowatt-hours annually, depending on the exact conditions.

A quick way to estimate your own kWh/kW/year number (which means the kWh output per the kW system capacity per year) is to start with your location. In the United States:

>> **Cloudier northern states** may be closer to 1,100–1,300 kWh/kW/year.

>> **Most of the Midwest and South** see 1,300–1,500 kWh/kW/year.

>> **Sunny western states** like Arizona or Nevada can reach 1,500–1,800 kWh/kW/year.

You don't have to guess, though. The National Renewable Energy Lab (NREL) has a free tool called PVWatts in which you can plug in your address, system size, tilt, and azimuth to see an estimate. That number you see — the kWh your system would produce for each installed kW per year — is the same number (or should be close to) the one that installers use in their proposals.

For example, a 4.75 kW DC system might generate around 6,800 to 7,000 kWh per year (4.75 × ~1,450), depending on tilt, azimuth, shading, and location. A larger 10 kW system, using the same assumptions, could produce 13,000 to 15,000 kWh annually.

Varying output expectations

Of course, the actual solar power production from any system doesn't come in even, monthly slices. Solar power follows the rhythm of the Sun: Longer days and higher Sun angles in summer mean more power generation, while shorter winter days naturally bring less. Rainy spells, snow cover, or a week of cloudy weather also show up on your bills.

One of the biggest surprises for new solar power system owners is that their production isn't the same every month. A rainy May or a snowy December will cut into power generation, while an unusually sunny season may bump it up. That's why annual totals are a better measure than any single month's output.

Making Quick Calculations on System Cost and Size

The title of this chapter is "Deciding Whether Solar Power Is Right for Your Home." The truth is, in most cases, the answer (without all the figuring in this chapter) is already *yes*. Solar power works; the technology is proven; and the benefits are real. The more useful question to ask is: What is the cost, and how do I benefit by using solar power?

That's where some quick back-of-the-envelope math comes in. You don't need to model every sunlight hour of the year or forecast utility rates 20 years out to get a feel for whether solar power versus investment makes sense. With just a few inputs, your system size, its expected output, and your utility's rates, you can sketch out the savings, compare them to upfront costs, and see how long it might take for the system to pay for itself.

This section walks you through those quick calculations and gives you the confidence to size up any solar proposal and know whether the numbers add up in your favor.

Calculating cost savings on your electricity bill

In Chapter 2, you can find out how to read your electricity bill and track your energy use. That knowledge is the foundation for figuring out how solar power can help you save on the related costs. Your bill tells you how many kilowatt-hours (kWh) you use and what you pay for them, and those are exactly the numbers you need to estimate savings from solar power.

The basic math follows these steps:

1. **Find your proposed solar power system's expected annual production (in kWh).**

   Find out how to estimate this in the section "Modeling your output" earlier in this chapter.

2. **Calculate your effective utility rate based on your electricity bills.**

   As described in Chapter 2, divide the total amount due on your electricity bill by the total kWh amount consumed for the month. Then take the average of this effective rate for 12 months.

3. **Multiply the number from Step 1 (your annual solar power output) by the number from Step 2 (your effective utility rate).**

   The result is the gross savings solar delivers in a year. Most people can use their average effective rate (see Step 2) for this calculation. That's the simplest way to get a quick estimate.

For example, suppose your system produces 10,000 kWh per year and your utility rate averages $0.20 per kWh. Multiply them together: 10,000 × $0.20 = $2,000 in annual savings. If your household only uses 8,500 kWh in a year, then your savings will be capped at offsetting that usage (unless your utility pays you well for the extra). See some sample annual savings calculations in Table 4-3.

Simple Annual Savings Calculation

Estimated Annual Solar Production (kWh)	Utility Rate ($/kWh)	Annual Savings ($)
8,500	0.20	$1,700
10,000	0.20	$2,000
10,000	0.27	$2,700

REMEMBER

If you're on a time-of-use (TOU) plan, your rate changes depending on the time of day. That makes the math trickier, because solar panels tend to produce during the cheaper daytime hours while you might be buying expensive electricity in the evening. For quick savings estimates, stick with your average rate. When you get a detailed proposal from an installer, they'll model the TOU curve for you.

This quick math won't capture every wrinkle of your rate plan or seasonal variation, but it gets you in the ballpark. From here, you can compare those savings to system cost and figure out how long it will take to break even on your investment.

Putting the numbers together: Figuring payback

After you know how much money your solar power system can potentially save each year, the next question is how long the system will take to pay for itself. This length of time is called the payback period. Think of it like buying a fuel-efficient car: You spend more upfront, but you eventually make up the difference through lower gasoline bills. With solar power, the savings come from avoided electricity costs.

Follow these steps to quickly figure your payback period:

1. **Take the installed system cost from your solar system proposal.**

2. **From that installed cost, subtract any government incentives that you can take.**

    For example, you may be eligible for a federal tax credit or state rebate.

3. **Divide the number resulting from Step 2 by your annual savings.**

    You can find instructions for figuring annual savings in the preceding section.

These steps give you the number of years that elapse before the system pays itself off. After that, the electricity it generates is essentially free, aside from minor maintenance costs.

For example, suppose that your 10 kW system costs $22,500 upfront. You qualify for the 30 percent federal tax credit, reducing the net cost to $15,750. From your earlier calculation (see the preceding section), you know that the system will save you about $2,000 a year. Divide $15,750 by $2,000, and you get a payback period of just under 8 years.

After the payback period, the system is like a paid-off car — but one that keeps generating value. With panels lasting 25 years or more, that means 15+ years of "free" electricity after the system has paid for itself.

REMEMBER

Payback is a helpful benchmark, but it's not the whole story. Solar power systems also give you a shield against rising utility rates, add value to your home, and reduce your reliance on the grid.

Chapter 5

Making Your Home More Energy Efficient

When most people think about investing in solar power, their minds jump straight to installing panels on the roof. But before you start sizing a system or running payback numbers (see Chapter 4), take a step back and really consider your energy use. The cheapest kilowatt-hour of energy you'll ever find is the one you don't have to use in the first place. In other words, cutting down on wasted electrical use in your home is the fastest, simplest way to lower your bills and shrink the size of the solar power system you may later decide you need.

Back in the early days of residential solar power, experts often talked about *negawatts* — the watts of energy you save instead of generate. The concept of cutting down on energy use still holds true today. Every lightbulb you swap for a more efficient one, every drafty window you seal, and every smart thermostat setting you dial in reduces your demand on the power grid. And that attention to the details of your energy use means your eventual solar installation can be smaller, cheaper, and better matched to your lifestyle.

This chapter explores practical steps you can take inside and outside your home to boost energy efficiency and try out solar-powered devices beyond the installation of panels on the roof. Some efficiency measures involve simple habits, such as shading a window or venting your attic. Other measures are creative, even playful, with projects such as solar fountains or landscape lighting. Each creative project gives you a taste of solar power's versatility and moves you closer to a cleaner, more cost-effective home.

Knowing Why You Look to Energy Efficiency First

Solar panels are exciting, but they shouldn't be your first step for saving money on your electricity bill. Think of your home like a leaky bucket: You can pour in more water (by adding solar panels), or you can patch the leaks (by improving efficiency). Patching the leaks first makes everything else work better.

As noted in this book's first edition, efficiency is the lowest-hanging fruit of the energy world — and this description is still true. Every kilowatt-hour you don't use is a kilowatt-hour you don't have to either buy or generate, which means you'll need fewer panels on your roof, spend less money upfront, and see faster payback if you do invest in a solar power system. Efficiency also multiplies a solar power system's benefits: By lowering your demand, you can cover a higher percentage of your electricity use with the same number of panels.

REMEMBER

Efficiency improvements aren't just about cutting waste. They often make your home more comfortable and your environment healthier. Shading windows can reduce glare and keep rooms cooler. Sealing drafts can cut down on winter chills and the need to heat your house. Smart thermostats can balance comfort with savings. And many upgrades are surprisingly simple and inexpensive when compared to a major solar power system installation.

So, before you even start running solar system calculations (expected electricity bill reduction, the size of system needed, and a payback period), take stock of your home's energy use. Find out:

>> **Areas of energy inefficiency.** Ask yourself where you lose the most energy that you could be saving. For example, do your drafty, single-pane windows let the heat out? Do you still have lighting fixtures that use incandescent lightbulbs? Do your south- or west-facing windows let in sunlight that fights with your air conditioning?

>> **Which upgrades (that address inefficiencies) could shrink your bill right away.** Can you replace your drafty windows with tight-fitting, double-paned windows to decrease your heat loss? Can you add blinds to your south-facing windows to lessen the strain on your air conditioner? Can replacing your fixtures and/or incandescent bulbs with LED lighting make an impact in your energy use?

By asking and answering these questions first, you set yourself up for a smoother, smarter solar power journey. Think of it this way: Your home is an ecosystem. The way the sunlight hits your windows, how your attic traps or releases heat, the type of lights you switch on at night, and even the way you use water all interact to shape your energy bill. You don't need to tackle everything at once. By making your living spaces just a little smarter and more efficient, you'll often see savings show up on your electricity bill right away.

A smaller, well-matched solar power system usually saves more money than an oversized one fighting against an energy-inefficient home.

Starting Inside the House

When it comes to energy efficiency, the easiest wins are usually found indoors. That's because the bulk of your energy use comes from things you rely on every day: heating and cooling, lighting, hot water, and appliances. Small changes in these areas add up quickly, and unlike a major solar installation, many of them cost little or nothing to try.

As the book's first edition pointed out, "energy efficiency starts at home with the things you can see and touch every day." This observation remains true, but today you have more tools and technologies than ever before to use for addressing efficiency. From reflective window films to ENERGY STAR–rated appliances and LED lighting, modern upgrades can cut wasted energy dramatically without changing your lifestyle.

This section focuses on the everyday places inside your home where you can make efficiency improvements that prepare the way for investing in solar power. Some improvements are about comfort, keeping living spaces cooler or warmer without extra electricity. Others are about convenience, like using solar-powered technology to replace technology that you plug in to charge with grid power. Together, these upgrades and improvements give you a head start on lowering costs before your first solar panel is even installed.

Sheltering living spaces from the Sun

Direct sunlight streaming into your home can feel wonderful on a cold winter day, but in the summer, it's often your biggest energy enemy. As the book's first edition noted, "sunlight is a double-edged sword — it brightens your rooms but can also drive up your cooling bills." Those sunbeams carry heat, and when they pour through your windows unchecked, they turn your living room into a greenhouse. That means your air conditioner has to work harder and longer, driving up your electricity use right when rates are highest.

The good news? Controlling sunlight doesn't require major construction, and you have options for exercising control:

TIP

>> **Install blinds, curtains, or other window coverings.** Having these sunlight blockers during the hottest part of the day can lower indoor temperatures by several degrees. If you prefer natural light, reflective window films or semi-transparent blinds are another option. They block a large portion of infrared heat while letting most visible light pass through, so your space stays bright but cool.

For renters or budget-conscious homeowners, even removable reflective films or insulated curtains can deliver a noticeable improvement.

>> **Plant shade trees outside of south- and west-facing windows (for the northern hemisphere) or north- and west-facing (for the southern hemisphere).** This strategy can be one of the most effective for reducing energy use long-term. A deciduous tree provides shade in the summer and drops its leaves in the winter, letting the low Sun help warm your home. Over time, that single landscaping choice can save as much cooling energy as a high-efficiency appliance upgrade.

>> **Modern windows offer even more help for saving on energy use.** Double- and triple-pane glass with low-emissivity ("low-E") coatings reflect infrared heat while maintaining visibility. If you're building or remodeling, investing in efficient windows pays dividends in both cooling and heating seasons.

Table 5-1 gives you a quick look at the potential impact on cooling of efficient shading window treatments.

Shading Strategy	Cooling Load Reduction	Cost Level
Closing blinds/curtains	5–10 percent	Low
Reflective window film	10–15 percent	Low-Medium
Insulated/thermal curtains	10–20 percent	Medium
Shade trees (south/west)	15–30 percent	Medium-High
Low-E double-triple pane windows	20–40 percent	High

Shading isn't just about comfort. By keeping indoor temperatures naturally lower, you reduce the number of hours your AC compressor has to run. That not only saves electricity but also extends the lifespan of your cooling system. Over the years, that reduced wear and tear can mean fewer costly repairs or replacements.

Close blinds and curtains on south- and west-facing windows during the afternoon and then reopen them in the evening to release heat and let cooler air in.

Venting your attic and cooling off the entire house

If you've ever climbed into your attic on a hot summer day, you know how extreme the temperatures can get; they're often 30 to 40 degrees hotter than outside temperatures. All that heat doesn't just stay up in the attic. It radiates downward into your living spaces, forcing your air conditioner to work harder. In fact, if left unchecked, attic heat gain can add hundreds of dollars a year to cooling bills.

Proper attic ventilation acts like a pressure relief valve. By letting hot air escape and cooler outside air flow in, you reduce the thermal load pressing down on your ceilings. That reduction can mean the difference between your AC running constantly or cycling less often, which lowers both energy use and wear on your system.

One of the simplest solutions to attic heat retention is a ridge vent or soffit vent system (see Figure 5-1). These passive vents use natural convection: Hot air rises and exits through the ridge while cooler air enters through soffit vents along the eaves. For homes in very hot climates, an attic fan, either electric or solar-powered, can actively pull hot air toward the fan and push it out through the vent. Solar attic fans, in particular, are attractive because they only run when the Sun is beating down hardest (and when the attic needs cooling the most).

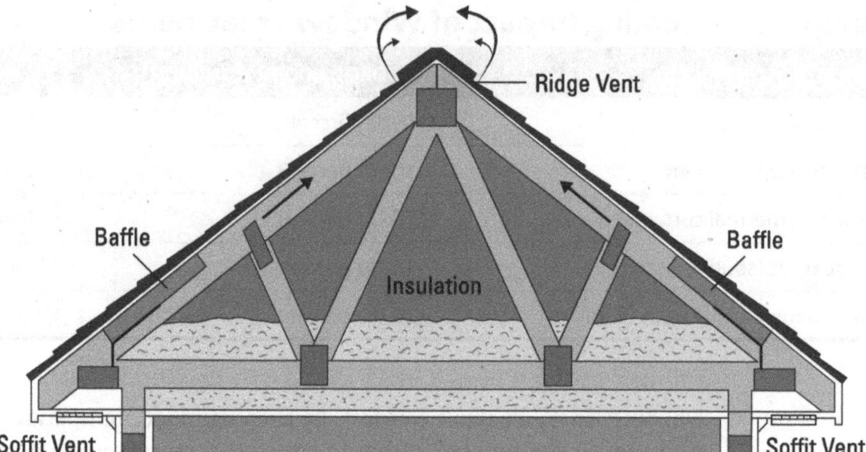

REMEMBER

Attic ventilation works best when paired with proper insulation. Insulation slows the transfer of heat from your attic into the rooms below, while ventilation reduces the overall heat load. Together, they create a one-two punch that can dramatically lower cooling costs and improve comfort. Check out Table 5-2 for a look at potential load reduction for various cooling strategies.

TABLE 5-2   **Attic Cooling Strategies**

Cooling Strategy	Cooling Load Reduction	Cost Level
Ridge + soffit vents	10–15 percent	Low-Medium
Solar attic fan	15–25 percent	Medium
Electric attic fan	20–30 percent	Medium-High
Adding insulation	20–40 percent	Medium-High

TIP

In hotter climates, really consider installing a solar attic fan. It's quiet, maintenance-free, and runs hardest when the Sun is at its peak.

Providing an endless source of purified drinking water

Energy efficiency isn't just about cooling and lighting; for houses that use well water or another independent water treatment system, it also touches the way you use water. Clean drinking water usually comes at the cost of pumping, filtering, or other treatment, and often plastic bottles (if you're buying drinking water at the store). Solar-powered water purification systems give you another option: fresh, safe drinking water created right at home without adding to your energy bill.

At their simplest, water purification devices use the Sun's heat to distill water. A solar still is a basic example: Sunlight heats water inside a transparent cover, the water evaporates, and then it condenses on the cover to drip into a collection container. The result is purified water, free of most contaminants. While small stills don't produce large volumes, they're effective demonstrations of how the Sun can replace electricity or fuel in an essential process.

Modern versions of water purification systems go further. Portable solar-powered purifiers combine photovoltaic panels with pumps and filters, allowing you to clean water anywhere you have sunlight. They're especially useful for camping, off-grid cabins, or as backup during emergencies when power and tap water may be unavailable.

Beyond the novelty, solar-powered water purification reinforces a larger concept: Your home can provide essential resources independently of the power grid. Whether you're topping off water bottles in the backyard or storing clean water during a power outage, these systems show how solar power can deliver reliability and peace of mind in everyday life.

Reading under the Sun at night

One of the easiest ways to cut your electricity use is by swapping out old lighting. Traditional incandescent bulbs convert most of their energy into heat rather than light, making them hugely inefficient. Compact fluorescents (CFLs) improved efficiency years ago, but today's LED bulbs are the clear winner — they use up to 85 percent less electricity than incandescent bulbs and last 20 to 25 times longer.

Now combine that lightbulb efficiency with solar power, and you can light your spaces at night using the energy you captured during the day. Solar-powered lamps and lanterns store sunlight in small rechargeable batteries and then provide hours of light after dark. Indoors, they're handy for reading or task lighting without plugging into the grid. Outdoors, they're perfect for patios, porches, or even campsites.

The best part about solar lighting is how accessible it is. You don't need an electrician, a roof mount, or even much space. Place the lamp by a sunny window during the day or use outdoor solar fixtures that charge in direct sunlight. At night, you'll have free light — proof that even small-scale solar can reduce reliance on the grid.

Every watt counts. Replacing a handful of old bulbs with LEDs, or adding a solar lamp to your routine, is a small change that adds up to significant savings over the years.

Improving efficiency with everyday appliances

After heating and cooling, appliances are the next biggest driver of household electricity use. And they're also one of the easiest places to make meaningful improvements. Unlike insulation or window upgrades, appliance efficiency upgrades are familiar, incremental, and often happen naturally as old equipment wears out.

In Chapter 2, you discover how appliances contribute to your overall energy usage. Here, the focus shifts from measuring that usage to reducing it before you decide on size for a solar power system. Every kilowatt-hour you don't need is one you don't have to generate, store, or pay for.

Modern high-efficiency appliances can use dramatically less electricity than older models. Refrigerators built today use a fraction of the energy of models from the 1990s. High-efficiency washing machines reduce both electricity and hot water use. Heat pump dryers and water heaters cut energy demand even further by moving heat instead of creating it. These upgrades quietly shrink your home's energy footprint every hour of every day.

ENERGY STAR certified appliances provide a reliable shortcut when comparing options for replacing older appliances. The label doesn't mean top of the line, but it does mean that the appliance meets strict efficiency benchmarks without sacrificing performance. Over the lifetime of the appliance, savings on utility bills often exceed the upfront cost difference.

Some appliance upgrades that matter the most include the

>> **Refrigerator/freezer.** These cooling devices run all day, every day, making them a constant energy draw. It's worth prioritizing replacement if your existing fridge is 10–15 years old or if you have a rarely used older second fridge in the garage or basement.

>> **Water heater.** One of the largest household energy users, water heaters cycle daily whether you notice it or not. Upgrading makes sense if the heater is older or if your household has high hot-water demand.

>> **Clothes dryer.** This appliance Uses short but very high-power bursts. It's worth attention if you do frequent laundry and are open to changing habits or upgrading to a more efficient model.

>> **HVAC equipment.** Your heating and cooling equipment is often the single biggest contributor to electricity use. Prioritize upgrades if the system is aging, oversized, or inefficient. Making changes here typically overlaps with comfort considerations and insulation improvements.

>> **Dishwasher.** Most homes use the dishwasher frequently, and efficiency between models varies widely. Newer units can significantly reduce both electricity and water use, especially in busy households.

Moving to Outside the House

After you tackle the easy energy-efficiency wins inside your home, it's time to step outside. Your yard, garden, and outdoor living spaces may not show up directly on your electricity bill the way your refrigerator or air conditioner does, but they're still full of opportunities to use energy more wisely. In fact, outdoor upgrades often combine efficiency with aesthetics, convenience, and even fun.

Think about all the ways you use energy outdoors: lighting pathways, running pumps for fountains or pools, keeping lawns green, or creating welcoming spaces to gather at night. You can power every one of those uses by capturing the energy from the Sun. Solar-powered outdoor products are some of the most accessible and affordable technologies on the market. You can pick them up at a hardware store, set them in place, and start saving instantly, no wiring or electrician required.

This section looks at practical and creative ways to extend solar power's reach beyond your rooftop. From lighting your yard to pumping water with solar fountains, you'll see how easily you can make your outdoor spaces more efficient, sustainable, and enjoyable.

Lighting your yard with solar energy

Yard lighting is one of the easiest and most satisfying ways to bring solar power outside. Instead of running electrical wiring through your landscaping or paying extra to keep grid-powered lights on all night, you can have solar fixtures do the work for you. They soak up sunlight during the day and automatically turn on after dark, lighting paths, gardens, patios, and entryways without adding a cent to your electric bill.

Modern solar yard lights have come a long way from the dim, short-lived versions of years past. Today's models use high-efficiency LEDs and lithium-ion or lithium-iron-phosphate batteries, giving you brighter light and longer runtimes. Many include dusk-to-dawn sensors that turn lights on automatically at sunset and off at sunrise, so you don't have to remember to do it. Table 5-3 offers a look at some common types of solar lighting for your yard.

Common Types of Solar Yard Lighting

Type	Best For
Pathway lights	Walkways, driveways
Spotlights	Highlighting trees, architecture
Wall-mounted sconces	Entryways, patios
String lights	Entertaining spaces, pergolas
Floodlights	Security lighting

Installing solar yard lighting is as simple as placing the fixtures where they'll get several hours of direct sunlight during the day. For best results (even double the performance), place fixtures thoughtfully; keep them angled to collect the sunlight and clear of overhanging trees or rooflines that might cast shade.

Solar yard lighting isn't just about aesthetics; it's efficient. By switching outdoor lighting to solar, you reduce nighttime electricity demand and avoid the cost and hassle of trenching or wiring. It's also a perfect entry point if you're new to solar power because it's low risk, low maintenance, and instantly impactful.

Literally going green with landscaping

Landscaping isn't just about curb appeal, but it's one of the most powerful tools you have to naturally control the temperature around your home. Strategic plantings can reduce heat gain in the summer, shelter your house from cold winter winds, and even help direct rainwater more efficiently. Done right, a few well-placed trees or vines can lower your energy bills as effectively as a new appliance upgrade.

The key to making landscaping effective is to work with the Sun's path, not against it.

>> **Deciduous trees,** the kind that lose their leaves in winter, are perfect for planting on the south and west sides of your home. In summer, their leafy

canopies block the harshest sunlight, shading walls, windows, and even roofs. In winter, when their branches are bare, they let the lower-angle sunlight stream through, naturally warming your home.

>> **Evergreen trees and dense shrubs,** on the other hand, are ideal for the north and northwest sides. They act as windbreaks, slowing cold winter gusts that can sneak through cracks and under eaves.

>> **Vines and trellises** are another clever way to green your exterior. Fast-growing vines can create living "green walls" that shade exterior surfaces and cool the air around them through *evapotranspiration,* a natural cooling process in which plants release moisture into the air as water evaporates from leaves and soil. A vine-covered trellis over a patio or window can reduce indoor temperatures while adding a beautiful, natural look.

>> **Ground cover matters, too.** Lawns and low-lying plants reflect and absorb heat differently than bare soil or concrete. Replacing heat-absorbing pavement with permeable surfaces and greenery can reduce the heat island effect immediately around your home, keeping the area cooler in hot weather.

Table 5-4 offers some energy efficient landscaping strategies, mainly for homes in the northern hemisphere. If you're in the southern hemisphere, you'll want to reverse the best placement referenced below.

TABLE 5-4   **Landscaping Strategies for Energy Efficiency**

Feature	Best Placement	Primary Benefit
Deciduous shade trees	South and west	Block summer sunlight
Evergreen windbreaks	North and northwest	Block winter winds
Vines on trellises	South and west walls	Shade exterior surfaces
Ground cover & permeable surfaces	Around paved areas	Reduce heat island effect

REMEMBER

Landscaping takes time to mature, but the energy savings grow along with your plants. A well-planned yard can provide decades of passive cooling and wind protection.

Warming up the pool water with solar power

If you have a swimming pool, you may know how much energy you need to keep the water warm — and how chilly that first dip can be without a heater. Solar pool heating is one of the most cost-effective and straightforward ways to harness the Sun's energy. It doesn't require tying into your home's electrical system, and once installed, a solar heater can extend your swimming season by weeks (or even months) each year without raising your energy bill.

The basic idea is simple: the solar heater pumps pool water through a series of dark-colored, Sun-facing collector panels (usually made of durable plastic or rubber). As the water circulates through the collectors, it absorbs heat gathered from the Sun before flowing back into the pool. Because pool pumps are already part of most systems, many solar pool heaters (see Figure 5-2) don't need extra equipment, they can tie into your existing plumbing.

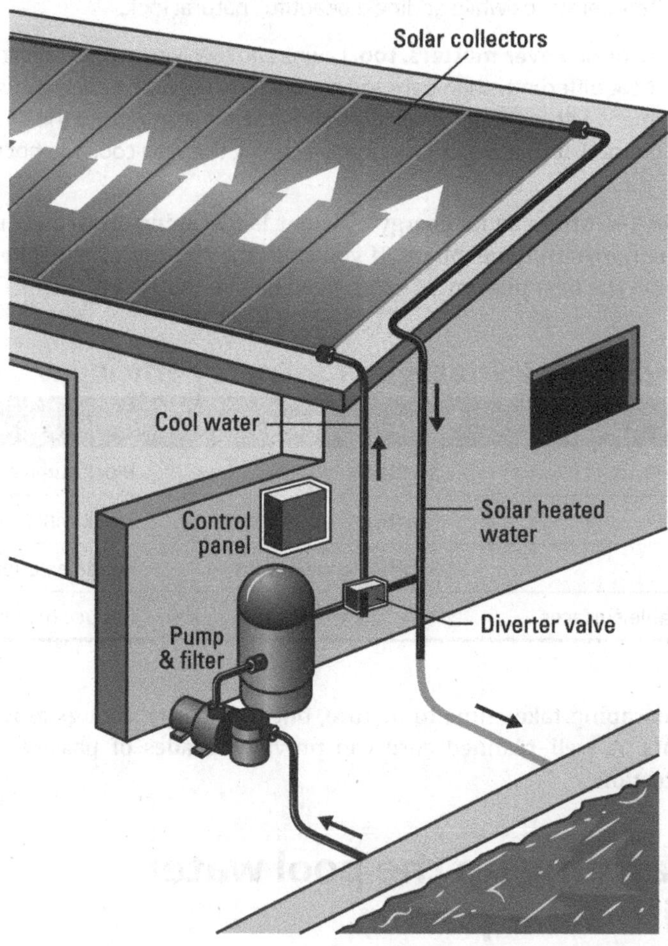

FIGURE 5-2: Diagram of a solar pool heating system.

Some systems are fully off-grid, using small solar-powered pumps to circulate water during the day. Others piggyback on your pool's existing circulation pump. Either way, the principle is the same: Let the Sun do the heating and save your utility costs for something more fun: like pool floats.

Pumping water to new heights with a solar fountain

Few solar gadgets bring a yard to life quite like a fountain powered by the Sun. There's something satisfying about watching water arc through the air or cascade down a tiered basin, all without plugging in anything. Solar-powered fountains are simple, elegant, and a great way to add visual interest and soothing sound to your outdoor spaces while showcasing solar energy in action.

The setup is straightforward. A small solar panel, usually mounted nearby or floating on the water's surface, powers a low-voltage pump. When sunlight hits the panel, the pump kicks on, circulating water through a nozzle or fountain head. Figure 5-3 illustrates the system components. There's no wiring, no trenching, and no ongoing electricity cost.

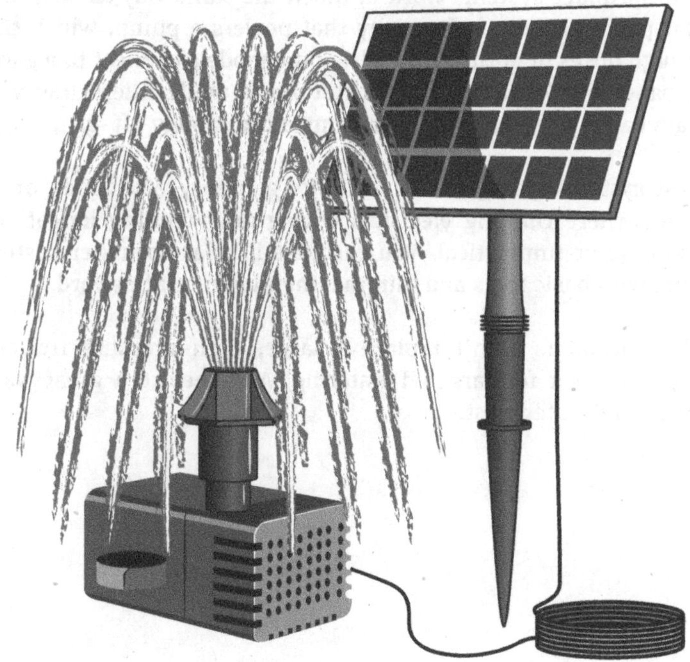

FIGURE 5-3: A solar panel powering a small pump that circulates water through a decorative fountain.

Some solar fountains come as all-in-one kits, drop them in a birdbath or basin, and they start running in seconds. Others let you pair a separate panel with a pump, which gives you more flexibility to place the panel in full sunlight even if the fountain itself sits in the shade. Many kits can be installed in less than an hour by using basic hand tools.

Because they only run when the Sun is shining (unless equipped with a small battery), solar fountains are low-maintenance and energy-free to operate. They can even help aerate ponds or birdbaths, improving water quality while adding a decorative element.

Getting creative with a solar-powered sprinkler

If you love a lush lawn or thriving garden, sprinklers are a summertime staple, but they can use a surprising amount of energy when paired with electric pumps. Solar-powered sprinklers offer a simple, eco-friendly alternative. By pairing a compact photovoltaic panel with a low-voltage pump, you can irrigate your yard or garden using nothing but the power of the Sun.

Solar sprinkler systems work in much the same way as solar fountains. A small solar panel generates electricity that powers a pump, which then pushes water through hoses or sprinkler heads. Some models connect to a garden hose and use the panel to power a timer or booster pump, while others draw water directly from a rain barrel or storage tank, creating a completely off-grid irrigation loop.

These systems are especially useful for gardens, raised beds, or remote corners of a yard where running electrical wiring or extending irrigation lines would be expensive or impractical. You can install solar sprinkler systems in just a few hours with basic tools and minimal maintenance afterward.

Solar sprinklers won't replace a large, high-pressure irrigation system, but they're perfect for targeted watering. And they're a great way to reduce both energy and water waste.

Chapter **6**

Getting the Gist of How Installation Works

After you explore the reasons for and benefits of using a solar power system, you can shift your thinking from theory to reality. Installing solar power involves money, equipment, paperwork, timelines, and people. Understanding how all those pieces fit together will make you a much savvier solar homeowner.

Whether you're weighing your investment options, deciding between DIY and hiring a contractor, or simply wanting to know what to expect after you sign a system proposal on the dotted line, this chapter gives you the inside scoop. You find out how to make smart financial choices, navigate the installation process step by step, and work confidently with codes, inspectors, and contractors to bring your solar project to life.

Paying Attention to Your Pocketbook

Solar power isn't just an environmental choice; it's a financial one. For most homeowners, the decision to go solar begins and ends with the bottom line: How much will it cost, how much will it save, and when will it pay for itself? Understanding the basic cost structure of a solar power project is the first step toward making smart decisions.

The total price of a residential PV system depends on several factors, including system size, equipment type, roof characteristics, local labor rates, and incentives. You find that

>> Most systems are priced by the watt, with installed costs in the United States typically ranging between $2.50 and $4.00 per watt (before incentives). A typical 6 kW system might cost between $15,000 and $24,000 up front, but that's not the full story.

>> Federal tax credits, state and local rebates, and utility incentives can dramatically reduce the total system cost. If you installed a solar power system in 2025, the federal Investment Tax Credit (ITC) covers 30 percent of eligible costs in the form of a tax credit, which means a $20,000 system could have its net cost reduced by $6,000. Because this tax credit expired on December 31, 2025, if you plan to install a system in 2026 or later, you may still indirectly benefit from some incentives — for example, by entering into a third-party ownership model, such as a solar lease.

REMEMBER

Back when residential solar power systems were first gaining traction as a supplement to power from the grid, homeowners often faced higher prices and fewer incentives. Systems were smaller; credits were less predictable; and the payback horizon was longer. Today, the financial landscape is far more favorable. Equipment costs have plummeted and financing options have expanded, making solar one of the most compelling home investments available. Table 6-1 gives you a look at costs for installing a solar power system at the time of this writing.

TABLE 6-1   **Typical Installed Solar Costs (2025)**

System Size	Typical Range	Gross Cost @ $3/W
4 kW	Small home	$12,000
6 kW	Average home	$18,000
10 kW	Large home/EV owner	$30,000

TIP

Be sure to subtract the amount of incentives from the gross sticker price to find your net cost. This net figure is what you use to calculate your return on investment (see Chapter 4).

Beyond incentives, it's crucial to factor in long-term savings. Most solar systems produce reliable energy for 25 years or more, and utility electricity rates generally rise over time. The savings accumulate gradually but powerfully. A homeowner who offsets most of their usage can easily save tens of thousands of dollars over the life of the system, depending on local rates and escalation trends.

In the past, many solar buyers focused mainly on upfront price, sometimes at the expense of system quality or performance. Today, smart solar shoppers look at lifetime value. That perspective includes choosing equipment that will last, installation practices that minimize maintenance, and financing that doesn't erode savings through high interest rates or *escalators* (contract provisions that allow payments to change over time based on triggers such as inflation or cost of living).

Considering the Best Overall Investments

When you're planning a solar project, focusing on the panels themselves is easy to do, but smart solar investments look at the big picture. The best returns often come from prioritizing energy use efficiency and smart home upgrades first — then designing your solar system around your optimized energy use. Spending a bit of time on strategic home efficiency improvements (see Chapter 5) before solar roof system installation can reduce the size and cost of your system without sacrificing comfort or performance.

Think through and plan for overall cost effectiveness:

>> **Overall home energy efficiency.** The less energy you use, the fewer solar panels you need to offset that usage, and the lower your upfront costs. Energy efficiency measures like upgrading insulation, sealing ducts, replacing outdated appliances, and switching to LED lighting can shrink your electricity demand significantly. These kinds of improvements often pay for themselves quickly, sometimes faster than the solar installation itself. (See Chapter 5.)

>> **Component quality and durability.** In addition to system efficiency, think about durability and component quality. The panels and inverters you choose (see Chapter 3 for an overview of system components) will be on your roof for decades, so cutting corners to save a few dollars upfront can cost more in the long run. Higher-quality components often come with longer warranties and better performance over time, which translates into greater energy yield and fewer maintenance headaches.

WARNING

>> **Financing structures available.** Many homeowners finance solar power systems through loans. And loans can be a great way to spread out costs, but high-interest loans or those with escalating payments can eat into savings that you realize from your system. Look for straightforward loan terms without ballooning rates or complicated escalators. Pay cash if you can or find low-interest financing to keep more of the savings in your pocket.

Some solar financing offers are designed to look affordable upfront but become much more expensive over time. Be cautious of loans with balloon payments, introductory interest rates that reset after a few years, or escalating payment schedules that rise faster than your utility savings. Sales pitches that focus only on the first-year monthly payment can be especially misleading. Always ask for a complete payment schedule showing how much you'll pay over the life of the loan, and if the terms aren't clear, keep shopping.

REMEMBER

Buying the cheapest solar power system isn't always the best investment. A slightly higher upfront cost for quality equipment and smart prep work can yield far greater returns over the life of your system.

Getting the Most Out of Your Equipment

Your solar panels may be the star of the show, but they're only as good as the *supporting cast* (the other system components — see Chapter 3). The overall performance and reliability of your system depend on how well your equipment works together and the structural choices (such as panel tilt and orientation) you make before installation. A few smart decisions up front can help you squeeze every kilowatt-hour out of your system for decades to come. Table 6-2 offers an overview of typical component lifespans and warranties.

TABLE 6-2   **Typical Lifespans and Warranties**

Component	Typical Lifespan	Typical Warranty
Panels	25–20 years	25-year performance, 10–15 year product
Inverters	10–15 years	10–15 years
Racking & mounting	25+ years	20–25 years
Wiring & connectors	25+ years	Varies
Monitoring equipment	10–20 years	Tied to inverter warranty typically

A solar power system — like a chain — is only as strong as its weakest component. Choosing high-quality, well-matched equipment sets you up for decades of dependable performance.

Starting with panel quality

Not all solar panels are created equal. Reputable manufacturers back their products with 25-year performance warranties, guaranteeing that the panels will still produce a high percentage of their original output after decades on your roof.

Cheaper panels may save a few dollars upfront but often degrade more quickly, which can cost more in lost power production over time. Look for strong product and performance warranties, reliable manufacturers with great reviews, and panels that have been tested to withstand your region's climate conditions.

Choosing an inverter wisely

The inverter is the workhorse of your system, converting DC electricity from the panels into AC power your home can use. Inverters typically have shorter lifespans than panels, often 10 to 15 years, so it's smart to factor in a replacement or extended warranties when planning your investment. A well-chosen inverter keeps performance high and maintenance low over the system's lifetime.

I cover various inverter configurations (string inverters, microinverters, and DC optimizers) in Chapter 4, so I don't rehash those options here. The key at this stage is to choose a reliable product from a reputable manufacturer and make sure it's properly matched to your system's size, shading, and layout.

Connecting and monitoring your system

Mounting hardware, wiring, and monitoring systems are also essential pieces of the solar power puzzle. Quality racking minimizes roof penetrations and keeps panels secure through wind, snow, and extreme heat. Investing in reliable wiring and connectors prevents small electrical losses and reduces the likelihood of future failures. Modern systems often include online monitoring platforms that let you track production in real time, flag issues early, and share data with your installer for remote troubleshooting.

Installing a Photovoltaic System

After you correctly size your system (see Chapter 4), choose your equipment (see Chapter 3), and figure out your financing (see Chapter 7), you can actually plan to get the panels on your roof. Installation involves a mix of planning, paperwork, construction, and coordination. Even when you hire a contractor to do the heavy lifting, understanding the basic steps of system installation helps you set realistic expectations and keep the project on track.

Installing in a nutshell

Every solar installation follows the same basic roadmap, whether it's a small DIY project or a full-service professional job. The total timeline can vary dramatically, from as little as three weeks in fast jurisdictions to six months or more where permitting or utility approvals move slowly. Knowing the process steps and sequence upfront will help you keep things moving and avoid unpleasant surprises.

Follow these steps for unfolding a typical project:

1. **Assess the installation site and prepare a system design.**

   Your installer (or you, if you're going DIY) evaluates the roof structure, electrical panel capacity, shading, and site orientation and then prepares a system design. For most contractors, this step involves a combination of on-site inspection and drone or satellite imagery.

2. **Submit building permit and interconnection applications.**

   Local building permits and utility interconnection applications are necessary documentation that verify your permission to construct your solar power system and connect it to the utility grid. Some cities issue solar permits the same day; others take several weeks and may involve multiple rounds of review.

   Many times, contractors can handle applying for building and interconnection permits. To find out about this process in your area, visit the website of your local government authority (usually your city or county) for permit instructions.

3. **After permits are obtained (or at least applications are submitted), order your solar power system equipment.**

   You need to order the panels, inverters, racking, and other necessary system components or have them pulled from inventory, if they're in stock. Some installers pre-stock, while others wait for permit approval to avoid storage costs.

4. **With permits in place and equipment onsite, install the solar power system.**

   Mounting hardware goes in first, followed by panels, wiring, and inverters. For a typical residential system, this step takes 1–3 days, depending on roof complexity and system size.

5. **Arrange for code inspection and utility sign-off.**

   After the solar power system is installed, a local building inspector reviews it for code compliance. After passing inspection, the utility conducts a final interconnection review and issues Permission to Operate (PTO). Only then can your system legally export energy to the grid.

Table 6-3 gives you a look at a typical timeline for residential solar power system installation.

TABLE 6-3   **Typical Residential Solar Timeline**

Step	Typical Duration	What's Happening	Who's Involved
Site assessment & design	1–2 weeks	Roof/electrical evaluation, final design	Homeowner, installer
Permitting & interconnection	2–8+ weeks	Local permit and utility application	Installer, authority having jurisdiction (AHJ), utility
Equipment procurement	1–3 weeks	Panels and inverters ordered	Installer (or you)
Installation	1–3 days	Mounting, wiring, inverter	Installer (or you)
Inspection and PTO	1–4 weeks	Building and utility sign-offs	Inspector and utility

Deciding between a contractor or DIY

In the early days of residential solar, many homeowners installed systems themselves, but as regulations have grown more complex, going DIY is far less common today. Most grid-tied systems must comply with stringent building, fire, and electrical codes, plus utility interconnection rules that often require licensed work. For most homeowners, hiring a qualified solar contractor offers a smoother, faster, and safer path to an approved installation.

That said, DIY can still make sense in certain scenarios: off-grid systems, ground mounts, or small plug-and-play kits. The tradeoff is clear: lower upfront costs in exchange for more dedicated personal time, elbow grease, responsibility, and risk. Table 6-4 offers you a quick look at some factors of installation and how they differ with a contractor or DIY approach.

     Contractor versus DIY Installation

Factor	Contractor	DIY
Permitting and paperwork	Handles this for you	You handle it all.
Code compliance	Professional expertise	You must learn and follow NEC/local building codes.
Timeline	Faster after scheduling	Dependent upon your availability
Upfront cost	Higher labor cost	Lower, but more personal time investment
Risk	Covered by warranties and insurance	You're liable for mistakes.

REMEMBER

DIY work can impact warranties and financing. Some lenders and equipment manufacturers require professional installation to maintain coverage or qualify for loans.

Working with kits

Pre-engineered solar kits can bridge the gap between full DIY and hiring a contractor. These kits include panels, racking, inverters (often microinverters), wiring, and sometimes even pre-labeled diagrams. They're especially common for ground mounts, carports, or accessory structures where roof work isn't required.

Modern kits have improved tremendously in ease of use, but they still require careful planning. Most grid-tied systems need electrical sign-off from a licensed contractor to pass inspection. Some homeowners use kits to handle mechanical installation while outsourcing the electrical portion, a cost-saving compromise that can work well if local regulations allow it.

Addressing the safety challenges

Solar installations involve working at heights and dealing with high-voltage electricity. Even experienced DIYers must treat these projects with caution and address these challenges:

>> **Safety-first work environment.** Fall protection, stable ladders or scaffolding, and weather awareness are essential when you're working on the roof. Think about renting professional equipment if you don't own industrial-quality ladders or scaffolding and fall-protection harnesses.

>> **Electrical safety operations.** On the electrical side, lockout/tagout procedures (like blocking electrical current, which prevents accidental contact with live electricity), proper grounding, and compliance with National Electrical Code (NEC) rapid-shutdown rules are critical for safety when installing a solar power system.

Electrical faults, arc flashes, or water leaks can create real, life-threatening hazards if you're not installing systems correctly. If you're unsure about any step in the installation process, bring in a licensed electrician or contractor to handle it.

Being Realistic about Codes and Regulations

Installing a solar power system isn't just about mounting panels and flipping a switch; it's also about playing by the rules. Every solar project must meet a slew of building codes, electrical codes, fire safety standards, and utility interconnection requirements — all of which exist to ensure that systems are safe, reliable, and compatible with the grid. Navigating these rules can feel intimidating, and a basic understanding goes a long way toward avoiding delays and headaches.

Most *local jurisdictions* (your city or county that governs building projects) adopt versions of the National Electrical Code (NEC), which are updated every three years. NEC requirements cover everything from conductor sizing and grounding to rapid shutdown rules, which are designed to protect firefighters by ensuring rooftop circuits can be de-energized quickly in an emergency. Local building codes also regulate roof penetrations, setbacks, and load-bearing capacity. Fire codes may require clear access pathways on rooftops, especially in jurisdictions with strict firefighter access rules.

Utilities add another layer of regulation. Before you can legally connect to the grid, you must apply for interconnection approval, which involves submitting system designs, equipment specifications, and inspection documents. (See the section "Installing in a nutshell" earlier in the chapter.) Some utilities offer seamless online processes; others rely on paper forms and lengthy review periods. Understanding your utility's timeline and requirements upfront is key to setting realistic expectations for the timing and circumstances of your solar power system installation.

On the honor system: Choosing to follow code

In the past, some DIYers operated on the "honor system" — installing solar without obtaining permits and assuming that if the installation worked, it was fine. While that approach may have flown decades ago, it's a bad idea today.

Unpermitted systems can cause major problems down the line:

>> Utilities may refuse to interconnect with an unpermitted solar power system.

>> Home insurers may deny claims that result from an unpermitted solar installation.

>> Future buyers may balk at purchasing properties with noncompliant systems.

Many jurisdictions now use satellite imagery and aerial inspections to flag unpermitted solar installations, so "flying under the radar" with an unpermitted solar power system isn't really an option anymore. Even if you live in a remote area with limited enforcement, following code isn't just about compliance, it's about safety and system longevity. Proper grounding, conductor sizing, and mounting aren't bureaucratic red tape; they protect you, your home, and the grid.

Cutting corners on permitting can cost you more down the line. Unpermitted systems that are discovered often need to be retrofitted — at the homeowner's expense — before they can be legally interconnected or before the home that has the system can be sold.

Passing code

After a solar power system is installed, inspections ensure that everything meets the regulations of code and safety standards. This process usually involves two steps:

>> **A building or electrical inspection by the local authority having jurisdiction (AHJ).** Inspectors will check that wiring is properly sized and terminated, grounding is correct, conduit runs are neat, disconnects are labeled, and the system complies with setback and access rules.

>> **A utility inspection or paperwork review for interconnection approval.** Utilities then verify that the system matches the approved design and that all documentation is in order before issuing a Permission to Operate (PTO) verification.

Without PTO, you can't legally export electricity to the grid — even if your system is physically complete, working, and up to the AHJ's code.

Hiring a Contractor

For most homeowners, hiring a qualified contractor is the single most important decision in the solar power system installation process. A good contractor can make the experience smooth, efficient, and even enjoyable. A bad one can cause costly delays, code problems, or performance issues that linger for years. Fortunately, with a bit of preparation, you can separate the pros from the pretenders.

Knowing what a contractor can do

Solar contractors typically handle the full scope of your project: site assessment, system design, permitting, equipment procurement, installation, inspection coordination, and utility interconnection. Some even assist with financing or applying for rebates and incentives. One of the biggest advantages of hiring a contractor is that they know the local codes, utility quirks, and permitting processes inside and out. What might take a homeowner weeks to learn, they can often complete in days.

Look for contractors with relevant licenses (such as an electrical contractor's license or solar installer license, depending on your state), insurance (including general liability and workers' compensation coverage), and certifications (such as NABCEP — the North American Board of Certified Energy Practitioners). These are strong signals that they follow industry best practices.

Be sure to ask the contractors that you're interviewing whether their company uses in-house installers or subcontractors. In-house crews typically offer more consistency and accountability.

Getting bids

You should always solicit multiple bids before signing a contract for your solar power system. This ability to compare bids gives you a clearer picture of market pricing, helps reveal *outliers* (bids that are unusually high or suspiciously low), and gives you leverage when negotiating terms. Aim for at least three quotes from reputable companies.

Each bid, or quote, should include

>> **Total system cost, cost per watt, and payment schedule.** That is, the full installed price, how it breaks down on a per-watt basis, and when payments are due (including any deposits, milestones, or final payment.

>> **System size and estimated annual production.** The size should reflect your historical electricity use, available roof or site space, and realistic production assumptions for your location.

>> **Equipment details.** The bid must outline the number and cost of all components, including panels, inverters, racking, wiring, and any monitoring or battery equipment — all with clearly identify manufacturers and model numbers.

>> **Contractor's experience and references.** For example, how long they've been installing solar, how many systems they've completed, and recent customer references or case studies.

>> **Project timeline.** As much as possible, the contractor should estimate the timeline needed for the entire project. See the section "Installing in a nutshell" earlier in the chapter for the steps involved in a project.

>> **Warranty information.** Warranties include equipment warranties (for panels, inverters, and batteries, if applicable) and workmanship or installation warranties.

>> **Assumptions about incentives and financing.** The bid should clearly state which tax credits, rebates, or utility programs are assumed, and whether financing terms are estimates or guaranteed.

Comparing bids

After you have multiple bids in hand, compare apples to apples. Some contractors may propose slightly different system sizes or inverter setups, which can skew the total price. Focus on cost per watt, equipment specifications, and estimated annual production to get a fair comparison. Check out Table 6-5 for a sample bid comparison.

Watch out for vague bids that omit key details, such as any of the key details mentioned in the preceding section — that's often a red flag. Financing terms also deserve scrutiny. Some installers bundle loans with *escalators* (which can change payment terms) or prepayment penalties that reduce long-term savings. Ask for full amortization schedules if financing is included.

Sample Bid Comparison

Feature	Bid A	Bid B
System Size (kW)	6.0	6.2
Total Cost	$18,000	$17,600
Cost/Watt	$3.03	$2.84
Panel Brand	QCells	REC
Inverter	Enphase	Solar Edge
Est. Annual Production	9,200 kWh	9,000 kWh
Timeline	10 weeks	8 weeks
Warranty	25 years panel/15 years inverter	20 years panel/15 years inverter

REMEMBER

The lowest bid isn't always the best value. Pay close attention to equipment quality, timeline, and warranty terms, not just the bottom line of the initial cost. For example, in the Table 6-5 comparison, Bid B may show the lowest cost, but the system also has lower annual production and a shorter warranty period. Consider all these factors before accepting a bid.

Interviewing a contractor

Interviewing potential installers is as important as reviewing their bids. Ask about their experience with your roof type, how many installations they've completed locally, and how they handle permitting and inspections. Inquire about their warranty service process. For example, if something breaks in year five, who do you call, and how quickly will they respond?

It's also wise to ask for references from recent customers, preferably in your area. Talking to other homeowners about their experience can reveal how well the company communicates, handles issues, and meets timelines.

TIP

Trust your instincts. If a salesperson seems pushy, evasive, or unwilling to provide details, that's a sign to move on.

Committing to a contract

Before signing anything, read the contract carefully. It should clearly outline a detailed scope of work, pricing, payment schedule, timelines, equipment, and warranties. Make sure incentive assumptions are realistic and that you understand what happens if permitting or interconnection delays occur.

In some cases, homeowners sign contracts months before installation begins, so it's important that the agreement is clear and binding. Don't hesitate to have a trusted friend, attorney, or experienced homeowner review it with you.

Working with a contractor

After work on your solar power system installation begins, communication is key. A good contractor will keep you updated throughout the permitting, installation, and inspection phases. Don't be afraid to ask for status updates, copies of permits, or documentation of completed work.

Chapter 7

Paying for Solar at Home

Installing a solar power system at home isn't just an energy decision, it's also a financial one. The good news is that you can find more ways than ever to make solar power affordable. Among federal and state tax credits (many of which expired at the end of 2025), rebates, loans, and special financing programs, you can reduce your upfront cost and shorten your payback period.

This chapter walks you through the major types of financial support available for homeowners who want help with the cost of installing a solar power system. This support ranges from direct and indirect subsidies and tax incentives to smart borrowing and alternative financing. In this chapter, you find out how to research programs in your own state, compare loan types, and decide what makes the most sense for your home and budget. Whether you're paying cash, taking out a loan, or bundling solar into a mortgage, understanding the finance piece will help you get the most out of your investment and avoid leaving money on the table.

The Different Types of Subsidies

When people first hear that solar can "pay for itself," they often wonder whether that's too good to be true. It's true, but only because a mix of federal, state, and local incentives can help tip the scales in your favor. These subsidies exist because solar power systems don't just benefit individual homeowners; they strengthen the grid, reduce peak demand, and create local jobs. Government policymakers decided that encouraging more rooftop systems is a smart investment for everyone.

Subsidies come in many shapes and sizes. And the rules and programs vary widely depending on where you live. You find that:

>> Some subsidies give you money back soon after installation. For example, a homeowner in Texas might get a one-time incentive from their local co-op after installing a rooftop solar power system.

>> Other subsidies may help you reduce your tax bill, and a few help ensure that solar power adds value to your home without adding extra property taxes.

>> Many states and utilities also offer ongoing performance-based incentives, rewarding you over time for the electricity your system produces. For example, a homeowner in Minnesota might earn rebates for every kilowatt-hour of energy generated.

REMEMBER

In many cases, you can combine subsidy programs. The more you qualify for, the faster your payback period shrinks. Together, these incentives can offset a surprising amount of your project's cost, sometimes 40 to 50 percent when used together effectively. The key is knowing what subsidies are available before you start your solar installation project. Many programs require pre-approval, and some have limited budgets or annual enrollment windows. Doing your homework early will save you time, money, and missed opportunities later.

TIP

You can explore current subsidy offerings near you through the Database of State Incentives for Renewables and Efficiency (DSIRE) at www.dsireusa.org. It's the most complete and frequently updated source for federal, state, and utility-level programs.

Looking for rebates

Rebates are the most straightforward form of solar incentive: You buy a system, fill out some paperwork, and get money back. Rebates are like a refund from your utility, city, or state — instant gratification for choosing clean energy. At this point, the federal government focuses mainly on tax credits for commercial providers of solar power systems, but rebates to the system owners usually come from state or local programs designed to speed up solar adoption in specific regions.

In the early days of home solar power systems, rebates were often the biggest financial boost available. Some state programs paid out more than $4 per watt of installed capacity, enough to cover half the system's cost. Because prices for solar system installations dropped, rebates also dropped. But they remain an important piece of the financial puzzle, especially in locations where installation costs are higher.

Rebate structures vary from one state to another. California and New York historically led the way with desirable rebates, but newer programs that offer a mix of upfront rebates, bill credits, and add-on incentives tied to system performance are popping up in places like Illinois, Minnesota, and Florida. In some regions, rebates target specific customers, such as low-income households or homes located in disadvantaged energy communities designated under the Inflation Reduction Act (IRA). Those targeted incentives can be worth an extra 10 to 20 percent of project costs.

Rebates generally come in two main forms:

>> **Capacity-based rebates:** These pay a fixed amount for every watt or kilowatt you install. For instance, if your local utility offers $0.20 per watt, a 7-kW system would earn a $1,400 rebate after it passes inspection. The total payment doesn't depend on how much electricity you actually generate; it's purely based on system size.

>> **Performance-based rebates:** These programs, sometimes called *production-based incentives* (PBIs), reward you for every kilowatt-hour (kWh) your system produces over a set period — often five to ten years. These rebates tend to pay smaller amounts per kWh (for example, $.02/kWh), but because they're tied to real production, they encourage efficient design and ongoing maintenance.

Most rebate programs are funded on a first-come, first-served basis, meaning the pot of money runs out when the annual budget does. Check the program status before signing your contract; some utilities publish online trackers showing how much funding remains. Many also require pre-approval before installation begins, so don't wait until your panels are on the roof to apply.

Programs often limit rebates to approved installers and equipment. Your contractor should be familiar with the paperwork, but double-check that the products and certification numbers on your proposal match what's eligible for payment.

Even as rebates shrink, they remain a meaningful signal: If your utility or state still funds them, it's a sign that they're serious about accelerating clean-energy adoption. Acting early ensures you capture the best benefits before programs phase out or shift toward larger commercial installations.

Qualifying for tax credits

If rebates are the quick-win in the solar world, tax credits are the long game — the incentive that gives back when you file your tax return. A tax credit directly reduces the amount of income tax you owe, dollar for dollar. It's not a deduction (which simply reduces your taxable income); it's better. Tax credits come right off your total tax bill, meaning they can save you thousands of dollars in one stroke.

The federal government no longer offers an income tax credit (ITC) for homeowners (after December 31, 2025), but solar power systems under third-party ownership (leases or power purchase agreements) can still qualify for tax credits through 2027. Be aware that your state may have some tax relief for you. Verify whether your state offers a credit and the current credit rate before you file your state taxes. Many states, counties, and even utility districts offer additional credits that work to reduce your tax bill based on what you spend on a solar system. These amounts can range from small flat credits to programs worth several thousand dollars.

The DSIRE database that I mention in the section "The Different Types of Subsidies" earlier in this chapter includes access to information about state credits that can help trim the net cost of a solar system.

No property tax increases

Installing solar panels often raises your home's market value, but that doesn't mean your property taxes have to rise along with it. Many states and local governments have recognized that solar improvements are environmentally beneficial and have created property tax exemptions to encourage the adoption of home solar power systems. Over the life of your solar array, these exemptions can amount to hundreds or even thousands of dollars in avoided taxes, all while increasing your home's resale appeal. It's one of the quietest but most durable incentives for going solar.

In most cases, these exemptions work in one of two ways:

>> **Exemption from reassessment:** Your property's taxable value isn't increased because of the solar installation.

>> **Exemption from taxation:** The solar system's value itself is completely excluded from your property's assessed value.

This means that you can enjoy the financial benefit of a more valuable, energy-efficient home without a corresponding increase in your annual property tax bill.

For example, California excludes active solar energy systems from property tax reassessment through January 1, 2035, and Florida and Arizona both have permanent exemptions written into state law. Some states go even further: New York allows a 15-year property tax exemption for qualifying renewable systems, and Massachusetts gives local governments the option to extend full or partial tax relief to solar homeowners.

Even if your state doesn't automatically exempt solar from increasing your property taxes, check with your county assessor's office. Some cities and local jurisdictions offer their own renewable energy exemptions or abatements, especially for rooftop systems under a certain size.

Incentives for home-operated businesses

If you run a business from home — whether that's a full-time operation, a freelance studio, or even a small side hustle — installing a solar power system can deliver tax benefits that go beyond those available to typical homeowners. When part of your property is used for business purposes, you may be eligible to deduct a portion of your solar installation as a business expense in addition to claiming residential tax credits.

The book's previous edition notes that the Internal Revenue Service allows business owners to depreciate the portion of the system used for business under the Modified Accelerated Cost Recovery System (MACRS). This still holds true. You can depreciate the percentage of your system tied to business use — for instance, 20 percent if you regularly use one-fifth of your home for work — which can accelerate the return on your investment in the solar power system.

Some states and utilities offer extra grants or loans for small businesses that install renewable energy systems. If your home-based enterprise is registered as a business, you may qualify for those programs as well.

When you combine depreciation, tax credits, and avoided electricity costs, the payback period for a home-operated business can be substantially shorter than for a purely residential system.

Net metering

Net metering (a system of billing that credits the owners of solar power systems for excess electricity that their systems send to the grid) is one of the simplest, and most powerful, incentives for homeowners who go solar. This billing method effectively runs your meter backward when you produce more energy than you use.

When your solar panels generate more electricity than your home consumes, say, on a bright summer afternoon, that excess power may flow into the local utility grid. In return, your utility credits your account for the value of those exported kilowatt-hours (kWh). Later, when your panels aren't producing enough to meet your needs, like at night or during cloudy weather, you draw power from the grid and use up those credits.

The book's previous edition describes net metering as a "give-and-take" relationship with your utility: You lend the grid power when you have plenty and borrow it when you need it. That's still true, but the rules for how you're credited have evolved since then.

Traditional versus evolving net metering

Under traditional net metering, every kWh you send to the grid earns you a full retail-rate credit, the same price you pay for electricity when you use it. For example, if your utility charges 25 cents per kWh, each surplus kWh your panels export earns you a 25-cent credit. Those credits roll forward on your bill, helping offset energy you buy later. Over a full year, your account may even out to nearly zero if you generate about as much as you consume.

In recent years, utilities and regulators have begun revising or replacing traditional net metering programs with net billing at reduced rates. The main reason comes down to grid economics, as follows:

>> When rooftop solar owners export power during the day, that energy competes with wholesale prices that have dropped sharply because of the growing number of solar systems on the grid. At the same time, utilities still have to maintain wires, transformers, and backup power plants for when the sun isn't shining.

>> By crediting exports at the full retail rate, older net metering rules didn't always reflect these real maintenance costs, leading many states to adopt "net billing" or "time-of-export" models that value solar energy based on when it's delivered to the grid.

Maximizing your cost savings

If you're a customer with a solar power system, this shift in net metering programs means that *self-consumption* (using as much as possible of your power product yourself) is now more valuable than ever. Using your solar power directly (for example, by running appliances during the day or charging a home battery) can save more money because you avoid using extra grid power rather than exporting energy to the grid for a reduced credit. So, in places where net metering is reduced, self-consumption becomes the key to maximizing energy cost savings.

To find out whether you're eligible for traditional net metering or one of these updated programs, start with your state's public utilities commission (PUC) or visit the DSIRE website at www.dsireusa.org. Most states list which utilities still offer traditional net metering, which have moved to net billing, and what credit rates apply. Your installer can also help interpret your local tariff, which can be an especially important step if you're trying to lock in a program before it transitions to new rules.

Tax-deductible home equity loans

Even with today's incentives, a solar system can still cost tens of thousands of dollars. One of the most popular ways to finance that investment is through a home equity loan or home equity line of credit (HELOC). These loans use the value of your home as collateral and often provide lower interest rates than unsecured personal loans or credit cards.

Because you're borrowing against your home, the interest you pay on a qualified home equity loan may also be tax-deductible, but only if the funds are used to "buy, build, or substantially improve" the property that secures the loan. Installing a solar power system clearly fits that definition because it adds a long-term improvement that raises your home's value and energy performance.

TIP

Keep copies of your solar contract and proof of payment. If you ever claim the interest deduction, the IRS may request documentation showing that the funds went directly toward the solar power improvement.

The difference between a home equity loan and a HELOC comes down to flexibility.

>> A home equity loan gives you a lump-sum payment up front with fixed monthly installments.

>> A HELOC, by contrast, works more like a credit line that you can draw from as needed. This arrangement is helpful if your solar project involves multiple phases, such as adding a battery or an EV charger later.

Interest rates can vary widely depending on your credit score and the equity available in your home, but they're typically lower than solar-specific loans offered through installers. The book's previous edition states that home equity financing often carries half the interest rate of unsecured consumer loans; that remains roughly true today.

Just be sure you understand the risks you take when you go for a home equity loan or line of credit. Defaulting on either can put your house at risk, so borrow conservatively. The good news is that a solar installation adds real resale value, meaning the improvement generally outlives the loan.

Researching All the Subsidy Options

Knowing that rebates, credits, and incentives exist is only half the battle; the other half is figuring out which ones you actually qualify for and how to claim them. Solar incentives come from many places: the federal government, your state, your city or county, and sometimes even your electric utility or local co-op. Because each program has its own eligibility rules, deadlines, and paperwork, the best thing you can do before you sign a solar contract is research every available option.

You have these resources:

>> **A U.S. national database:** Database of State Incentives for Renewables and Efficiency (DSIRE), available at www.dsireusa.org. This site provides a state-by-state list of tax credits, property tax exemptions, rebates, and net metering policies. It's maintained by the North Carolina Clean Energy Technology Center and updated regularly. You can search by ZIP code to see exactly which programs apply to your home.

>> **Your utility company's website:** Most utilities now have renewable energy pages listing available rebates, net metering programs, or time-of-use rate plans for solar customers. Don't skip this step; in many cases, utilities require preapproval before you install solar power to qualify for certain incentives.

>> **Bundled incentives from your installer or lender:** If you're financing your solar system, your installer or lender may also bundle incentives directly into the proposal. Many installers use incentive databases or software that automatically pulls in all eligible programs based on your address. That's convenient, but don't rely on it blindly, double-check the details yourself to make sure they match what's available in your state.

Recognize that incentives can change quickly. A state program available today might close next quarter or reopen under new terms. By doing your homework early and saving essential information as screenshots or downloadable program details, you'll have proof of the incentives you qualified for at the time of installation.

Getting a Loan for Your Solar Power System

Even after rebates and credits, many homeowners still need to finance part of their solar project. Luckily, you can find a good selection of loan options — from traditional banks to specialized "green" lenders — designed to make solar ownership affordable and accessible.

You're essentially trading an upfront investment for a fixed monthly payment that may be lower than what you currently pay your utility. Over time, your energy savings offset the loan, and after you pay it off, the electricity your panels generate is effectively free.

Before you pick an option, take a look at the basic structure of solar loans. Table 7-1 gives a quick side-by-side comparison of the main ways homeowners finance a solar power system installation; I offer you additional details in the following sections.

TABLE 7-1   **Common Solar Financing Options**

Financing Path	Who Provides It	Typical Term	Ownership	Advantages	Things to Look Out For
Traditional Bank or Credit-Union Loan	Local or national banks	5–15 yrs.	You	Simple structure, full ownership, flexible pay-off	Rates depend on personal credit; unsecured loans have higher APR
Energy-Efficient Financing Program	State/federal partner banks (PACE, HomeStyle Energy)	10–25 yrs.	You	Low rates, property-based qualification	May complicate resale, limited by state laws
Energy-Efficient Mortgage (EEM)	Mortgage lenders (FHA/VA/Fannie Mae)	15–30 yrs.	You	Rolls solar into home loan, low monthly cost	Requires audit, best for refinance or home purchase
Home-Equity Loan, Home Equity Line of Credit (HELOC), Cash-Out Refinancing	Banks or mortgage lenders	10–30 yrs.	You	Lowest rates, potential interest deduction	Secured by home, slower closing process
Third-party lease or PPA	Solar installer or developer	15–25 yrs.	Provider	Little/no upfront cost, maintenance included	No tax credits, contract transfer on resale

Borrowing money the old-fashioned way

Your neighborhood bank or credit union may offer home-improvement or personal loans for renewable-energy projects. These types of loans tend to come with fixed rates and 5- to 15-year terms; they're ideal if you want full ownership of your system from day one. A typical example might look like a 10-year loan for a $25,000 total system cost at a 7% fixed interest rate, which will cost you about $290/month.

The main advantages of a home improvement or personal loan are simplicity and control. You receive all available tax credits, you own the equipment outright, and you can sell your home without transferring an outside contract.

The previous edition of this book notes that community banks can often beat national lenders on rates because they understand local property markets; that concept still holds true today.

The main drawback of a personal loan is that unsecured loans depend on your credit profile, and interest rates can be several points higher than home-equity or mortgage-backed options.

Using an energy-efficient financing program

Specialized programs created by federal and state agencies make solar power system installations more affordable for homeowners who have a good payment history but limited cash.

Here are a couple of examples:

>> **Fannie Mae's HomeStyle Energy mortgage** lets borrowers fold the cost of solar, storage, or efficiency upgrades into a new or refinanced mortgage, spreading payments over 15 to 30 years at mortgage-level interest rates. Similarly, FHA 203(k) and VA Energy Efficient Mortgage programs can finance solar installations as part of a home purchase.

>> **PACE (Property Assessed Clean Energy) financing** (at the state level) remains a popular choice. Instead of a conventional loan, repayment appears as a line on your property-tax bill, often over 15 to 25 years. Because qualification is based on property equity rather than credit score, PACE opened doors for many homeowners who couldn't get traditional loans.

PACE loans attach to the property, not the borrower. That means selling your home can require you to pay off the balance unless the buyer agrees to assume it, which is a situation that lenders increasingly restrict. Always check local disclosure rules before signing.

Pursuing energy-efficient mortgages

Energy-efficient mortgages (EEMs) reward upgrades that reduce long-term housing costs. Because a solar power system trims your future electric bills, lenders may allow you to borrow slightly more without raising your total debt-to-income ratio.

To qualify, you'll need an energy audit or modelled-savings report estimating how much your system will cut energy use. Lenders then fold those projected savings into your *loan underwriting process* (in which the lender thoroughly assesses the lending risk); essentially, the lenders may treat solar power systems as a predictable, value-adding improvement rather than an expense.

EEMs work best when you're already refinancing or purchasing a home, since you're consolidating everything into one low-rate mortgage. They're less ideal for existing loans if you'd have to restart a 30-year clock just to add a solar power installation.

Thinking about other mortgage options

If specialized programs don't fit, you can still fund a solar installation with a cash-out refinance or home-equity product. With mortgage rates that are often two to three points below unsecured loans, this approach spreads the cost over decades while keeping monthly outflow predictable.

Before you refinance, look at the remaining term and rate on your current mortgage. If you're sitting on a very low rate, adding a separate home-equity loan may be smarter than resetting your main mortgage.

Past editions of the book emphasized that a solar power system's added home value usually exceeds the remaining loan balance once the system is paid off; that situation still applies today. The system becomes a built-in asset, lowering your household energy expenses long after the financing ends.

Considering Alternative Financing and System Access

If buying a system with cash or a loan isn't the right fit, you still have viable ways to get a solar power installation without owning the equipment. For example,

>> Common third-party arrangements exists in which a company installs and maintains the system on your roof and you pay for the energy it produces. In a lease like this, you make a fixed monthly payment for the right to use the system's output.

>> In a power purchase agreement (PPA), you pay per kilowatt-hour for the electricity the system actually generates, usually at a rate set lower than your utility price.

In both cases, the third party owns the system and claims the tax benefits; you get little or no upfront cost, predictable bills, and professional maintenance baked in.

These agreements trade maximum lifetime savings for simplicity and immediate cash-flow relief. Because you're not the system owner, your monthly costs are more like a utility bill replacement than an investment that pays itself off. Terms typically run 15–25 years and may include small annual escalators. If you plan to move, check the contract's transfer options: Most allow a buyer to assume the agreement or let you buy out the system after a set number of years. *Note:* Clear transfer language helps avoid hiccups in a home sale.

If your roof isn't a good candidate (too shaded, tricky structure, or a condo/HOA restriction), community or shared-solar programs may let you subscribe to a slice of a larger off-site project and receive bill credits for your share of the system's energy production. These subscriptions often require minimal commitment and can be a practical path for renters or households that are planning a move. Savings vary by state rules and utility credit rates, but many programs target a 5–15% bill reduction without any on-site equipment.

Chapter **8**

Living with Solar

Getting a solar power system installed (probably) on the roof of your home is a big deal. But now what? In this chapter, I tell you about living with solar power and the basics of monitoring, maintaining, and even expanding the system. You find out how to check on system performance, evaluate your utility bills and savings, and make sure that your system keeps doing its job over the long haul. Performing just a small amount of maintenance can help your system run smoothly and last as long as possible. This is the fun part because you don't have much system upkeep, but a few things are worth keeping an eye on.

Living with solar isn't about constant management, it's about knowing what results to expect, how to respond if performance seems off, and how to get the most out of your system. Think of this chapter as your owner's manual for living the solar life: simple steps, smart habits, and a few ways to level up when you're ready.

Monitoring System Performance

After a solar power system is part of your home, the focus shifts from an installation project (see Chapter 6) into a lifestyle. You may notice yourself watching the weather more closely, checking your solar monitoring phone app to see how much energy your system produced, or spotting patterns in your electric bill that never caught your eye before.

And when your solar system is up and running, it's easy to forget it's even there. After all, it doesn't make any noise and usually just works on its own. But checking in on it regularly helps ensure everything is running correctly and gives you the chance to catch small issues before they become bigger problems.

Most systems come with monitoring software — usually provided by your inverter manufacturer or solar system installer — and having access to this software is important. The software is usually either a smartphone app that you download or a website you can log into. You can use the software to track daily energy production, notice any dips in output, and compare your system's performance across different seasons. Figure 8-1 shows an example monitoring dashboard.

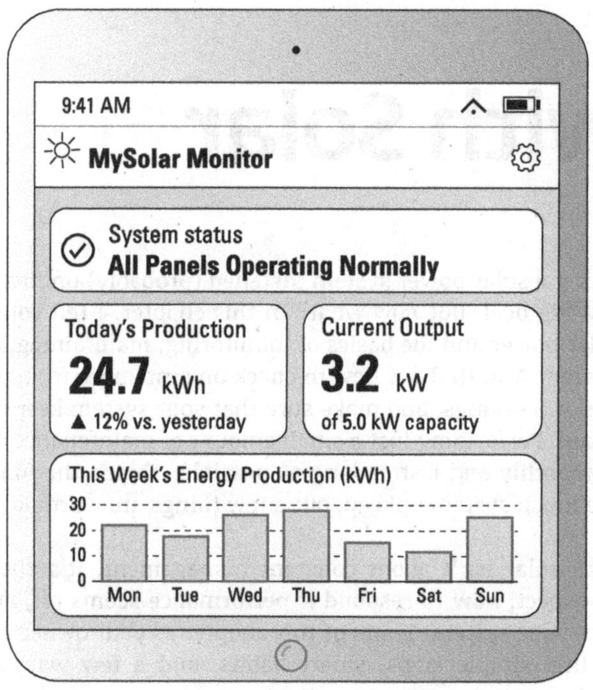

Knowing what to watch for

You don't need to check your solar power system or the monitoring app every day, but taking a look once a week, or even just once a month, is a good idea. (Although checking daily can be fun, too!) When you do visually inspect your system or check your app, look for these data and indicators:

>> **Consistent daily production:** Expect seasonal changes in output, but sudden or unexplained drops in energy production are a sign that something might be wrong or interfering with the system. Cloudy days will lower production, so a drop between 10–50 percent on cloudy days and 90 percent on rainy or stormy days is par for the course.

>> **Inverter status:** If your inverter (see Chapter 3 for description of an inverter and other components of your solar power system) has indicator lights that appear in your software or physically on the inverter unit itself, typically mounted on a wall near your electrical panel. (Most inverters are easy to see.) Check the indicators to make sure that they show operation as *normal* or *online*, for example. Lights that are red (instead of green) or error code messages — as you may expect — probably indicate a fault such as grid disconnection, communication error, or internal inverter issue.

>> **Shading or obstruction:** Obstacles such as leaves, snow, or new tree growth can block sunlight and reduce solar power output. Especially if your system is installed on your roof, be mindful about trees that might increasingly cast shade over time.

Take the little time necessary to monitor your solar power system. Even a small dip in energy production over time can add up to lost savings. Catching solar power system issues early will help ensure that you get maximum value from your system.

Warranty and performance guarantees

When looking at the energy output produced by your solar power system, keep your warranties and performance guarantees in mind. These elements can offer peace of mind if some indicator (such as unexpected drops in daily or monthly energy production) seems off in your monitoring app or if your system underperforms. Knowing what protections you have enables you to take action early and avoid the loss of potential savings.

Most solar power systems come with one of several types of warranties. If you aren't sure what warranties your system includes, you can start by calling your installer. Some companies offer a bundled "performance guarantee" that wraps multiple types of coverage into one. You can also look up your panel and inverter model numbers online to find information about the manufacturer warranties. Table 8-1 offers a look at some typical warranties and performance guarantees.

Warranties & Performance Guarantees for a Typical Solar System

Component	Typical Warranty	What It Covers
Solar Panels	20–25 years	Guarantees the panels' performance won't degrade beyond a certain rate over time and covers manufacturing defects that may occur
Inverter	10–15 years	Equipment failure for central or microinverter
Workmanship	5–10 years	Roof leaks or installation issues
Performance Guarantee	1–10 years	Installer "wrapping" — any kind of issue you encounter through guaranteeing a minimum energy production, usually annual

Because the performance guarantee is typically the shortest and easiest to measure, you should really pay attention to this aspect during your first few years of living with a solar power system. If your installer offered a performance guarantee, make sure you understand its definition — usually the guarantee has as a minimum annual production in kilowatt-hours. Also, know what happens if your system falls short of this guaranteed production. Because guarantees often last only 1 to 10 years, this window of time is when you'll want to monitor your system closely. Make sure to document your system's output and contact your installer promptly if the output seems off.

Tracking Billing and Savings

One of the best parts of installing a solar power system is seeing your electric bill shrink. And your utility will still send a monthly bill, but it may look a little different than before. The key for evaluating the bill is to know what to pay attention to and how to use your bill as a tool to confirm that your system is working as expected.

If your system is on a net metering plan (see Chapter 7 for information about this type of plan), you'll likely see line items for energy exported to the grid and credits applied against the electric power you used. If you're on a time-of-use (TOU) plan, the timing of your production and consumption matters, too. Chapter 2 helps you understand a TOU plan and your whole energy bill, but these ideas become real when post-solar billing shows savings or spots with warning signs.

Here are a few things to look for when reviewing your post-solar-installation bills:

>> **Lower charges overall:** In most cases, your monthly bill should be signifi-cantly lower than it was before your solar installation. Some months may even show near-zero balances.

>> **Seasonal swings:** Your savings will vary based on solar production, and your energy usage will vary, too. Depending on how much heating you use in cold months and air conditioning you use in hot months, you can expect your bill to increase or decrease. Both heating and air conditioning are energy intensive.

Your system's production can decrease as a result of cloud cover as well, so you can expect your bill to increase during extended cloudy periods.

>> **Unexpected increases:** If your bill suddenly rises, this could be a good sign that you need to check your monitoring app to confirm whether your system is underproducing (a decrease in generated power) or your household usage has changed (an increase in demand).

>> **Utility fees:** Even when you install a solar power system, you may still see fixed charges, minimum usage fees, or connection costs from your utility.

Instead of having to compare data between your utility bill and performance monitoring software platform, manually tracking some simple metrics in a spreadsheet or notes file may be a good idea. That way, you'll get the full picture in one place.

Table 8-2 gives you an example of tracking your energy usage in one place (a spreadsheet in this case).

TABLE 8-2 **Manually Tracking Your Performance in a Spreadsheet**

Month	Solar Production (kWh)	Utility Bill ($)	Notes
January	750	$18.45	Higher usage due to heating
February	780	$10.12	Sunny month, lower usage
March	610	$43.27	Low production and higher bill despite being sunny. Might be an issue.
...	...	...	...

Tracking your utility bills is one of the easiest ways to confirm whether your system is delivering savings. If the amounts look off, your monitoring data can help you dig deeper to find the root cause of system performance that doesn't match your expectations.

Performing Necessary Maintenance

Solar power systems generally require a low level of maintenance, and that's great! Because the solar panels themselves have no moving parts, systems can run for decades with little intervention. But "low-maintenance" doesn't mean "no maintenance," and a little effort goes a long way in keeping your system working efficiently.

Keeping solar panels clean

An effective way to maintain your solar system's expected output is also one of the simplest ways: Keep your panels clean. In many areas, regular rainfall is enough to rinse off dirt and dust. But if you live in a dry or dusty region, or notice buildup like bird droppings, pollen, or ash from wildfires, your panels may need occasional cleaning.

If you can do so safely (that is, you can reach the panels without climbing onto the roof or using unstable ladders), you can hose down the panels (usually from the ground) by using a gentle spray of water. You don't need chemical cleaners or pressure washers. Just be cautious if you have hard water because it can leave streaks or residue on the panels.

Hiring professionals to clean your panels is a smart option for solar systems installed on rooftops or in hard-to-reach places. Many companies now specialize in solar panel cleaning, and some even use drone technology to access and clean panels efficiently. Your installer may offer this service or be able to recommend a trusted provider.

Cleaning panels that have collected dust and debris can improve energy output by 5 to 15 percent, and sometimes even more, depending on conditions.

Watching for shading and physical damage

Over time, your solar system's environment can change. Trees grow, neighboring buildings go up, and seasonal shadows shift. All of these environmental changes can block sunlight and reduce your system's performance. You often see these issues physically before they show up in your monitoring software. Keep an eye out for overhanging branches or new obstructions that weren't obstacles when you first installed your system.

Debris is another occasional culprit of degraded solar energy output. Leaves, snow, or other buildup can lodge between or under panels. In some cases, animals such as birds or squirrels may try to nest under your panels. If you notice signs of debris or nesting activity, it's best to call a professional for a safe inspection and removal.

Also, when inspecting your panels, look for visible damage such as cracks or chips in the glass. This damage may result from falling tree branches or flying debris, but it might also point to manufacturing defects or issues related to the original installation. If you spot anything concerning, check your warranty to find out whether the damage you see might be covered.

Scheduling inverter checkups

Solar panels typically last 20 to 25 years or more, but inverters (another component of the solar power system noted in Chapter 3) usually have a shorter lifespan: about 10 to 15 years. This fact means that you'll likely need to replace your inverters at least once during the life of your solar power system.

Unlike panels, which gradually degrade, inverters often fail more suddenly. It's okay to wait until your inverter shows signs of failure; most homeowners replace theirs only when needed and not in advance. If you're a few years into your system, make note of when the warranty expires on your inverters and plan for possible replacement.

You can keep an eye on inverter health through your monitoring software (see the section "Monitoring System Performance" earlier in the chapter for more about this software). Most inverters also have indicator lights (either physically on the unit or in the app) that show whether everything's working as it should. A green or blue light usually means normal operation, while red lights or error codes may signal a fault.

If your inverter goes out, your solar power system will stop producing power until you replace that component. This situation can be frustrating, but replacement is usually straightforward. If your solar power system is still under warranty, contact your installer or the inverter manufacturer. If it's not under warranty, a local solar contractor can help you source and install a new unit.

Making an annual maintenance checklist

Once or twice a year, take 15 to 30 minutes to run through the quick maintenance checklist outlined in Table 8-3.

Task	What To Do
Visual panel inspection	Look for dirt, leaves, bird droppings, or physical damage
Inverter status check	Confirm status lights are normal, review any error codes if present
Monitoring app review	Check for consistent daily production and compare to previous months
Tree and shading review	Trim any overgrowth that could be casting new shadows
Warranty review	Make note of your warranties and performance guarantees applicability and expiration dates

Adding Storage or EV Charging

After your solar power system is up and running, you may start thinking about ways to get even more out of it. Two of the most popular upgrades are adding battery storage and installing an electric vehicle (EV) charger. Both additions can improve your energy independence and help you make the most of your solar power.

If you're planning a future upgrade to battery storage or EV charging, let your installer know early. The installer may be able to right-size your system or help you choose compatible equipment to make future add-ons easier.

Adding battery storage

Having a solar battery enables you to store excess energy generated by your solar power system during daylight hours for use at night or during power outages. If your utility company bills with time-of-use rates (see Chapter 2), using a battery to store electricity when rates are low can help lower your energy costs because you use that stored energy when rates are higher.

Most solar power systems have the batteries installed at the same time as the rest of the system, but you can also add on batteries later. If you're considering a system upgrade to include a battery, check with your installer about whether your current system is battery-ready. Some inverters are designed to support battery storage, but others may need extra components to do so. You can find out more about batteries in Chapter 9, but keeping this option in mind is important when you review your system's performance and electric bills.

Battery systems come with their own warranties, usually around 10 years or for a set number of charge cycles. Before reaching out to an installer, ask yourself a few key questions:

>> Do I experience frequent power outages that having stored battery power could help with?

>> Am I on a time-of-use rate plan in which electricity is more expensive in the evening?

>> Do I commonly produce more solar energy than I use during the day?

>> Can federal, state, or local incentives help offset the cost of adding battery storage?

If you answered *yes* to one or more of these questions, adding battery storage may make financial and practical sense. Frequent outages point to resilience benefits, time-of-use rates and excess daytime production suggest cost savings, and available incentives can improve payback.

Charging your electric vehicle with solar power

As electric vehicles (EVs) become more popular, they pair especially well with solar power generation. Charging your car by using energy from your own rooftop system can help reduce both your carbon footprint and your electricity bills.

If your home is on a net metering plan (as discussed in Chapter 7), charging your EV at night can still be solar-powered. The excess electricity your system generates during the day earns credits that can offset your evening use. If you're on a time-of-use rate plan, it may make more financial sense to charge during midday when rates are lower and your solar system is actively generating energy.

You can plug your car into a standard outlet for slow charging, or install a higher-voltage Level 2 charger (such as a 240-volt charger similar to what powers an electric dryer) to speed things up. Some advanced EV chargers can even integrate with your solar power system to optimize charging based on solar production and utility rates.

Chapter **9**

Going off the Grid

iving *off the grid* has an almost mythical appeal — a home that powers itself, a life free from utility bills, and the comfort of knowing you're insulated from blackouts or rising electricity prices. But the reality of going off-grid involves complexity and is often more demanding than most people expect. When you disconnect from your utility, you become your own utility. There's no backup power plant smoothing over cloudy days, no neighborhood transformer balancing your voltage, and no customer service line to call when your electricity supply falters. Every watt of energy you generate, store, and use is yours to manage.

Some homeowners choose off-grid living out of necessity: Remote locations where power lines don't reach or where the cost to extend service is extremely high are examples. In many rural areas, the price of a utility hookup can rival or exceed the cost of an entire solar-plus-battery power system, making an off-grid system the more economical choice. Other people are pursuing independence, sustainability, or resilience: the satisfaction of producing their own electricity and the peace of mind that comes from being unaffected by widespread utility power outages.

This chapter helps you understand whether going off-grid makes sense for your situation by walking through the structural, financial, and lifestyle implications of building your own stand-alone power system. You find out what equipment you must have, how it all works together, what it costs, and what an off-grid life actually looks like day to day.

Understanding Off-Grid Ramifications

Going off-grid isn't just about installing solar panels; it's about taking full responsibility for how you supply your home with electrical power. When you disconnect from your utility, you also disconnect from the enormous stability and flexibility that the grid quietly provides every day. Deciding to go off-grid means that you're not connected to a public utility for electrical power and that single choice transforms everything from the equipment you need to the way you use energy.

Choosing off-grid power has consequential trade-offs, including:

>> The off-grid equipment is more extensive than the kind that's typically used in a grid-tied solar installation.

>> The system must be sized for worst-case conditions: short winter days, long stretches of clouds, or unusually high household demand. That means batteries, charge controllers, backup generators, and monitoring systems all play major roles, not just the solar panels on the roof.

>> Because you no longer have the grid's stability to lean on, you need to think about energy differently. You may need to adjust how much electricity you use and when you use it, and you must realistically predict what output your system can provide.

>> Off-grid systems require diligence, regular maintenance, and an understanding of your own consumption patterns.

Figure 9-1 compares on- and off-grid solar power systems.

When off-grid power systems make sense

For some homeowners, off-grid living is a matter of necessity. Remote cabins, mountain homes, and rural properties often sit miles from the nearest utility line, and the cost to bring in grid service can be astronomical — sometimes tens of thousands of dollars per mile. In these cases, building your own power system is not just an alternative; it's the only practical option. For others, going off-grid is a deliberate choice rooted in independence, resilience, or environmental values. Producing your own electricity can feel empowering, especially when you're unaffected by blackouts or rising utility rates.

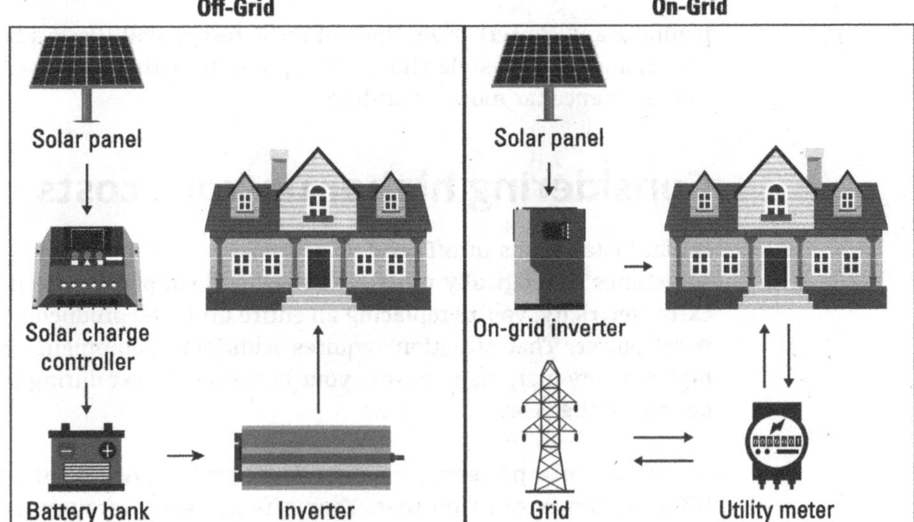

FIGURE 9-1:
Key architecture differences between on- and off-grid systems.

Taking accountability

But energy independence comes with accepting responsibility for your energy supply, your system, and its maintenance. Unlike a grid-tied solar system, which can lean on utility power at night or during storms, an off-grid system has to carry you through every scenario: sunny days, cloudy weeks, heatwaves, winters, and unexpected changes in power requirements.

REMEMBER

These situations mean you must install the system sized appropriately for the worst days, not the best ones. Your system needs enough solar capacity to meet your annual energy needs and enough battery storage to get you through dark stretches or surprise demand spikes. A grid-tied home might use a clothes dryer or induction cooktop without thinking twice; an off-grid home may need to schedule those energy usage loads for when the batteries are full.

When you install an off-grid solar PV system, the extra equipment you need is expensive, and it requires a lot of monitoring. Modern solar power systems have made active monitoring easier, but the core truth remains: Off-grid living demands active engagement. You must monitor your battery's state-of-charge, keep an eye on weather forecasts, and occasionally adjust your usage patterns to match the energy output that your system can deliver.

The shift to an off-grid system is not just technical; it's behavioral. People who thrive off-grid are usually willing to treat energy like a finite resource, similar to water pumped from a well or logs for a woodstove. Solar abundance in summer can feel limitless. But in winter, by contrast, getting adequate power requires planning and conservation. None of these responsibilities is a deal-breaker, but it does mandate a lifestyle choice. And going in with clear expectations will make the experience far more rewarding.

Considering higher upfront costs

Initial installations of off-grid solar systems cost more than grid-tied systems, sometimes dramatically more. The reason is simple: You're not just generating extra electricity; you're replacing an entire utility by producing all your own electrical power. That situation requires additional equipment, more storage, and higher safety margins to ensure your home has power during the lowest sunlight periods of the year.

A grid-tied solar power system can use the grid as a nearly infinite backup battery. Off-grid homes can't do that. Your off-grid solar array must be larger than a comparable grid-connected home installation, and your battery bank must cover more than the energy needed in a typical day. The battery backup must cover the darkest stretch of no-sun that you expect to experience.

REMEMBER

When you choose to install an off-grid system, you call the extra coverage required *oversizing*. When you install an off-grid solar PV system, you must oversize the system for the worst-case conditions. You don't get anywhere near 100 percent utilization under ideal solar power conditions (long hours of great sunshine on your system's panels) because you must oversize to account for the dark periods and the production difference between winter and summer.

Off-grid systems will require:

>> **Larger battery banks:** You need enough storage to power your home through nights, cloudy periods, and seasonal dips in production off-grid versus a typical smaller battery for grid-connected systems, which can rely on the utility as backup.

>> **Charge controllers:** These system components regulate the flow of electricity from the solar panels into the batteries to prevent overcharging or damage. This is a critical role when the batteries are your only energy reserve.

>> **System monitors:** The monitors must provide detailed, real-time information on battery state of charge, energy use, and solar production so you can adjust consumption and avoid running out of power.

>> **Backup generators (in most climates):** Because the supply of energy from the sun can be unpredictable for days or weeks during storms, winter months, or heavy cloud cover.

> **Additional safety disconnects and wiring:** You need these extras because you must be able to isolate batteries, solar arrays, and backup sources safely without relying on the utility grid to stabilize the system.

> **Energy-management hardware:** These hardware components in an off-grid system prioritize essential loads, shed noncritical circuits when power is low, and coordinate solar, battery, and generator operation to stretch limited energy supplies.

All of these extra system items add costs up front and over time. Batteries degrade and must eventually be replaced. Generators need fuel and periodic servicing. Solar panels can last decades, but in an off-grid environment, the supporting components (inverters, breakers, cabling) often see heavier duty cycles and may need replacement sooner than in a grid-tied system.

Noting the increased maintenance need

Solar power system maintenance is also a much bigger part of off-grid life. While today's lithium-iron-phosphate batteries have made off-grid systems far less hands-on than older lead-acid battery banks, you must still check the battery system status regularly. You must periodically clear snow, manage shading, reset breakers, and keep an eye on weather forecasts to decide whether to charge the battery with your generator before a long cloudy stretch comes along.

The key question is not whether off-grid systems work; they absolutely do, but whether you're comfortable with the costs, upkeep, and involvement that come with complete energy independence. Some homeowners enjoy being closely involved with the operation of their power system; others prefer letting the utility handle those responsibilities.

Looking green

A big appeal of off-grid living is the sense that you're doing something deeply sustainable, living closer to nature, generating your own clean energy, and reducing your dependence on fossil-fueled power plants. And in many ways, off-grid solar power is green: It runs on sunlight, produces no on-site emissions, and prevents costly, high-impact utility line extensions to remote homes.

But it's also important to recognize the full environmental picture. Off-grid systems require more equipment than grid-tied systems, especially larger battery banks and backup generators, and all that extra hardware has its own footprint.

>> You must replace batteries and generators periodically, and throwing them out can be costly and environmentally unfriendly. Modern lithium batteries last far longer than older lead-acid banks, but they still require mining, manufacturing, transport, and eventual recycling.

>> Even if you run your backup generators for only a few hours a year, they still burn fuel and require maintenance. Many off-grid homeowners choose propane as the fuel because it burns more cleanly; others opt for biodiesel blends to reduce carbon intensity. But unless you oversize your solar array and battery bank dramatically (which increases environmental impact), a generator is still part of most off-grid systems.

If you want the greenest off-grid system possible, invest in the most efficient appliances you can afford and temper your everyday energy use. Reducing your daily load means needing fewer batteries and far less generator runtime.

The good news is that off-grid systems can still be net-positive for the environment. Avoiding long-distance utility construction prevents land disturbance, and many off-grid households naturally become more energy-aware by upgrading appliances, reducing energy waste, and shifting large loads to sunny periods when solar energy production can cover them.

Surveying Off-Grid Solar Options

When you're planning to install and use an off-grid system, one of the biggest decisions you face involves how your system handles DC (direct current) and AC (alternating current) power. Solar panels naturally produce DC electricity, and batteries store it as DC. Your home, however, runs almost entirely on AC.

And so, off-grid solar power setups must juggle both current types, choosing when to keep electricity in DC for efficiency and when to convert it to AC so your house works normally. Older off-grid cabins may have stuck to simple DC lighting and water pumps, but modern off-grid homes combine the two current types so you can use refrigerators, microwaves, air conditioners, and power tools without rewiring your life. Consider these situations:

>> **Collecting and storing energy.** DC is the "charging side" of the system. It's the path energy takes from the panels that collect the solar energy into your batteries that store them without conversion losses. A DC-coupled charge controller sends clean DC power into your batteries with high efficiency.

>> **Using the energy for everyday living.** AC is the "living side" of the system that powers your outlets and appliances the way the grid would in a grid-tied system. The inverter then draws from those batteries to supply your home with AC, handling everything from your well pump's surge load to your nighttime lighting.

>> **Blending DC and AC.** Today's off-grid systems blend the two current types (DC and AC) seamlessly by using hybrid inverters that manage charging, switching, and power conversion behind the scenes. You don't have to choose one type or the other; you just need to understand how they work together so you can size the system correctly and avoid unnecessary energy losses.

REMEMBER

Because every off-grid system has to handle both types of electricity, it's useful to picture the energy flow. In earlier generations of off-grid technology, homeowners sometimes wired dedicated DC appliances, but modern hybrid systems have made full-AC homes not only possible but standard for comfort and convenience.

Understanding Batteries and Power Storage

When you go off-grid, batteries stop being optional accessories and become the beating heart of your entire electrical system. They determine how long your home can stay powered with back-up and how often you can expect to perform maintenance activities. Today, you usually have some form of lithium-ion batteries in your solar power system because this chemistry prevails in home battery systems, EVs, and electronics.

Within this family of batteries, lithium iron phosphate (LFP) has become the standard for off-grid homes because it's safe, stable, and delivers most of its rated capacity without stress. Earlier editions of this book warned that "batteries demand respect," and although lithium-ion has made off-grid living easier, that advice still applies. In any case, treat your batteries carefully and protect them from heat, cold, and overuse.

Lithium-ion batteries aren't your only option. For decades, flooded lead-acid batteries have been the workhorses of off-grid cabins, and sealed absorbent glass mat (AGM) batteries still show up in smaller systems. Nickel-iron batteries survive almost anything, but they can waste lots of energy. Each battery chemistry comes with a personality, and choosing wisely affects everything from comfort to cost.

Battery types

Table 9-1 offers a side-by-side look at the most common battery options:

TABLE 9-1   ## Common Off-Grid Battery Types

Battery Type	Advantages	Disadvantages	Best For
Flooded Lead-Acid (FLA)	Low-cost, well-understood, easy to source	Needs watering, must be vented, low usable capacity, short life	Seasonal cabins, tight budgets
Absorbent Glass Mat (AGM)/Sealed Lead-Acid	Reduced maintenance, spill-proof	Shorter life than lithium-ion, heavy	Occasional-use cabins
Lithium Iron Phosphate (LFP)	Long lifespan, high usable capacity, minimal maintenance, excellent efficiency, safer chemistry	Higher upfront cost	Most modern off-grid homes
Nickel-Iron (NiFe)	Extremely rugged, tolerates deep discharge	Very inefficient, bulky, expensive	Highly-niche setups

TIP

If you live off-grid full-time, choosing lithium-ion battery technology is almost always worth the extra upfront cost because it reduces both battery size and long-term replacement expenses.

Capacity and efficiency

A solar power system battery's capacity (measured in kilowatt-hours, or kWh) tells you how much energy the battery can store. Its usable capacity tells you how much of that energy you can actually use without shortening the battery's lifespan. The usable capacity matters enormously in an off-grid system in which you depend on your stored energy every single night. Check out Table 9-2 for a look at usable capacity.

TABLE 9-2   ## Usable Capacity of Battery Types

Battery Type	Cycle Life
Flooded Lead-Acid (FLA)	50%
Absorbent Glass Mat (AGM)/Sealed Lead-Acid	60%
Lithium Iron Phosphate (LFP)	90–95%
Nickel-Iron (NiFe)	70–80%, but with low efficiency

The chemical makeup of various batteries gives you choices and trade-offs. For example, nickel-iron batteries allow you to deeply discharge the batteries without harm, which gives them a relatively high usable percentage. The catch is that their round-trip efficiency is much lower than other chemistries, meaning you must generate more solar energy to refill them.

Efficiency determines how much of your solar energy you actually get back after storing it. Table 9-3 gives you an overview of battery efficiency by type. And efficiency is why lithium-ion (especially LFP) has become the default for modern off-grid solar power systems. Batteries with this makeup waste very little in the process of storing and providing energy. Lead-acid chemistries lose noticeably more energy, and nickel-iron loses the most, even though they are extraordinarily durable.

TABLE 9-3

Round Trip Efficiency of Battery Types

Battery Type	Cycle Life
Flooded Lead-Acid (FLA)	70–80%
Absorbent Glass Mat (AGM)/Sealed Lead-Acid	80–85%
Lithium Iron Phosphate (LFP)	95–98%
Nickel-Iron (NiFe)	60–70%

Maintaining batteries

In older off-grid solar power systems, battery maintenance became a hobby all by itself. Flooded lead-acid batteries needed regular watering, corrosion checks, equalization charges, and carefully managed ventilation. Many off-grid homeowners became experts at topping up cells with distilled water and cleaning terminals with baking soda, whether they wanted to or not.

Modern systems have different maintenance needs. Lithium-ion batteries, especially LFP, require almost no user maintenance. You have no requirements for watering, equalization, corrosion cleanup, nor gas venting. The built-in battery management system (BMS) quietly handles voltage balancing, high- or low-temperature protection, and charge control. For most homeowners, maintenance involves little more than glancing at the system monitoring app a few times a month.

Where lithium does require care is temperature. Lithium-ion batteries don't like charging below freezing and may reduce output at very high temperatures. If you live in a cold climate, your installer may use a heated enclosure or insulated battery cabinet. Hot-weather users may need shade, airflow, or a temperature-controlled utility space.

Lead-acid batteries remain more demanding. You must water flooded batteries periodically because charging causes water in the electrolyte to evaporate and break down over time. Good ventilation is mandatory because batteries release hydrogen gas while charging. You need equalization charges to control over-charging and prevent sulfate buildup, and you must check corrosion around terminals regularly. AGM and sealed lead-acid batteries reduce some of these chores, but they still need periodic voltage checks and can suffer a shortened lifespan if kept in hot environments.

Battery lifetime expectations

A battery's lifetime is typically measured in cycles: one full charge and one full discharge. Off-grid solar power systems will likely cycle batteries every single day, so lifespan isn't an abstract performance metric; it's a concrete prediction of how many years you'll go before replacing one of the most expensive parts of your system.

Different battery chemistries wear out at dramatically different rates.

>> **Flooded lead-acid (FLA) batteries** may lose significant capacity after just a few years of daily cycling. AGM batteries last somewhat longer but still can't compete with modern lithium-ion chemistries. Earlier editions of this book noted that deep discharging "punishes" lead-acid batteries, and that remains true: The deeper you drain them, the fewer cycles you get.

>> **Lithium iron phosphate (LFP)** batteries — the most widely used lithium-ion battery for off-grid systems — routinely lasts 6,000 to 10,000 cycles, depending on depth of discharge (the percentage change of discharge) and temperature; enough for more than a decade of daily use.

>> **Nickel-iron batteries** are almost a category of their own. They can last for decades, literally, but their drawbacks (low efficiency, high cost, and bulk) mean they're rarely a practical choice for modern off-grid homes. Still, it's not unheard of to find NiFe banks chugging along after 30 or 40 years, even in harsh environments.

See Table 9-4 for a quick look at battery lifetimes.

Lifetime of Battery Types

Battery Type	Cycle Life
Flooded Lead-Acid (FLA)	500–1500 cycles
Absorbent Glass Mat (AGM)/Sealed Lead-Acid	600–10,000 cycles
Lithium Iron Phosphate (LFP)	6,000–10,000 cycles
Nickel-Iron (NiFe)	10,000+ cycles

REMEMBER

Cycle life is only part of the equation when you think about what type of batteries you want to use in your off-grid solar power system. Temperature of the environment, depth of discharge, and charging practices all shape how long a battery actually lasts. Lithium-ion batteries prefer shallow cycling (20 percent depth of discharge) and moderate temperatures (15–30 degrees Celsius); lead-acid batteries prefer partial charging and can suffer if left at a low state of charge. The old rule still applies across all chemistries: "Take care of your batteries, and they'll take care of you."

Risks involved with batteries

Even though batteries are safer and more reliable than ever before, going off-grid means understanding the few ways things can go wrong and how to avoid them. Earlier editions of this book warned that batteries "are friendly servants but terrible masters," and that advice still holds. Each chemistry has its own personality, and knowing those quirks helps you design a system that's safe, predictable, and long-lasting.

>> **Flooded lead-acid batteries come with the most traditional risks.** As they charge, they release hydrogen gas, which must be vented so it can't build up and ignite. They also give off tiny amounts of acid mist, which corrodes metal components over time and can irritate skin or eyes if handled carelessly. Corrosion on terminals is common and increases resistance, which reduces performance.

Lead-acid batteries are also extremely sensitive to how deeply and how often they're discharged. Letting them sit at a low state of charge or draining them too far dramatically shortens their lifespan. These aren't dangerous problems, but they can be costly if ignored.

>> **Lithium-ion batteries, especially LFP, remove almost all of these old-school risks, but they introduce new ones.** Instead of producing gases or requiring watering, lithium-ion depends on its battery management system (BMS) to keep things safe. The BMS monitors cell temperature, voltage, and

charge levels; if anything drifts out of a safe range, the BMS simply turns the battery off. The battery shutoff is intentional, but it can surprise off-grid homeowners the first cold morning their battery refuses to charge.

Temperature extremes are the biggest concern with lithium-ion. These batteries cannot charge below freezing, and very high heat accelerates cell aging. Proper insulation, ventilation, or climate-controlled enclosures prevent nearly all of these issues.

Most safety problems associated with lithium-ion come not from the chemistry but from poor-quality manufacturing, which is why buying from reputable suppliers is so important.

>> **Nickel-iron batteries are unbelievably rugged, nearly immune to deep discharge, but they have their own operational quirks.** Their self-discharge rate is high, meaning they lose energy steadily even when you're not using them, and they also release small amounts of hydrogen gas that require ventilation.

The nickel-iron battery's low efficiency means that they waste a significant portion of the energy you put into them. This isn't a fire risk or safety hazard; it's an "annoyance risk" that forces you to oversize your solar array to compensate.

Specifying and Pricing Your System

You find out how to size a grid-tied solar power system in Chapter 4, and the goal is simple: estimate your annual electricity use and see how many kilowatt-hours your panels could replace. Going off the grid requires a totally different mindset when you consider the size and composition of your solar power system. Instead of thinking in annual terms, your concerns now hinge on daily use, worst-case conditions, and how many days of power you need to have stored when the sun disappears. Your solar power system can no longer fall back on the utility grid — it becomes the grid.

That's why off-grid system design starts with a bottoms-up load analysis, not an energy-use offset target. Then you consider which type of batteries can offer the best performance and which other components (solar panels, inverters, and so on) can make your system top-notch.

>> **For your bottoms-up, energy-need analysis,** you

- List every appliance, how many watts it draws, and how many hours per day it runs. This gives you a realistic daily energy budget, which is the backbone of the entire design.

- Recognize small loads like laptops and LED bulbs, which can add up gradually.

- Concentrate on large loads such as well pumps, freezers, electric ranges, and mini-split heat pumps can dominate your entire system. Many off-grid systems replace electric appliances with propane or super-efficient alternatives not because it's fashionable, but because it keeps the solar and battery system affordable.

» **After the energy-need analysis, you must right-size the battery bank —** something grid-tied homeowners barely think about.

- Off-grid homes need enough usable storage to power your house through each night plus however many cloudy days you want to ride out. While grid-tied systems essentially need "zero days" of autonomy, off-grid systems often plan for one to three days, depending on climate and comfort level.

- If you use 10 kWh/day and want two days of autonomy, you need 20 kWh of usable storage, which — depending on the chemistry of the battery backup system — might require a 22 kWh lithium-ion battery or a 40 kWh lead-acid bank.

» **After you choose batteries, consider the solar panels.** Grid-tied sizing for solar panels uses annual averages, but off-grid sizing must account for your worst solar month, usually December or January (in the northern hemisphere).

A system that looks fine in summer may fall dramatically short in winter. This situation is why many off-grid homes have arrays that look oversized to grid-tied homeowners: They're sized to the darkest weeks of the year so that you don't need to run a generator every morning.

» **In addition to solar panels,** your inverter also matters more in an off-grid system. Instead of simply converting DC to AC, the inverter must handle your home's peak loads: the highest wattage you might draw at a single moment.

Having a microwave and a well pump turn on at the same time can overload a small inverter, so off-grid designs often slightly oversize the inverter to avoid (colloquially) *blackouts by blender.*

Pricing an off-grid system is also different from pricing a grid-tied system. Batteries, enclosures, generators, charge controllers, balance-of-system components, and seasonal oversizing add up. (See the earlier section "Understanding Off-Grid Ramifications" for a look a comparison of on- and off-grid system components.) A compact, part-time cabin solar power system might cost under $10,000, while a full-time off-grid system — designed for year-round comfort — can easily fall between $25,000 and $60,000 depending on climate, chemistry, and expectations.

Peeking In on a Real-Life Scenario

Seeing an actual off-grid design in action helps make sense of all the individual pieces. In this example, you can take a look at a small full-time, off-grid home in a mild climate. The homeowner cooks electrically, works from home, and uses efficient appliances. It's a great case study because it sits at the intersection of comfort, reliability, and realistic budgets. I simplified everything in this example so that you can see the logical thought process without needing specialized modeling software.

The first step for designing your off-grid solar power system is to estimate daily energy use. In this example, the homeowner checks appliance labels, tracks usage patterns, and builds a rough usage tally, as depicted in Table 9-5.

TABLE 9-5

Example Daily Energy Use

Load	Daily kWh
Fridge	1.2
Lights	0.4
Laptop + Wi-Fi	0.3
Fans	0.6
Washer	0.3
Well pump	0.4
Cooking (induction)	1.0
Phantom loads	0.2
Total	4.4 kWh/day

This household averages about 4 to 5 kWh/day, which is typical for an efficient off-grid setup. But off-grid design relies on worst-case conditions, not averages. In winter, the sun is weaker, days are shorter, and clouds linger longer, so the system needs enough stored energy to ride through those stretches comfortably.

Don't design an off-grid solar power system for sunny June afternoons; design it for the darkest, shortest winter days.

A common design target for off-grid systems involves three days of *autonomy*, meaning the home can run for three full days without direct solar power production. Figuring an off-grid system design for this target follows these steps:

1. **Figure the required usable battery capacity for three days of autonomy from direct solar power.**

   - 4.4 kWh/day × 3 days = 13.2 kWh usable battery capacity

   - For lithium iron phosphate (LFP) batteries, which provide about 90 percent usable capacity, that translates to 13.2 kWh ÷ 0.9 ≈ 15 kWh of total installed battery storage

2. **Use the site's winter peak sun hours (3.5 hours/day) to calculate the daily winter energy production needs.**

   The system must generate enough energy to cover daily needs plus recharge one day's share of battery capacity:

   - Daily use: 4.4 kWh

   - Battery refill (⅓ of the total installed battery storage): ≈ 4.9 kWh

   - Total winter production needed (daily use + battery refill): 4.4 kWh + 4.9 kWh ≈ 9.3 kWh/day

3. **Calculate the minimum output of the solar array based on the winter production need.**

   9.3 kWh/day ÷ 3.5 h/day ≈ 2.7 kW solar array minimum production

4. **Round up the calculation from Step 3 to 3.0 to 3.5 kW for comfort and reliability.**

With the calculations complete, Table 9-6 shows what the finished system looks like.

TABLE 9-6

Example Off-Grid System Design

Component	Specification
Solar array	~3.2 kW
Battery bank	~15 kWh LFP
Inverter/charger	4–6 kW
Backup generator	3–4 kW

This setup provides a comfortable margin for year-round use. Summer brings abundant surplus energy; winter is steady but disciplined; and the generator rarely runs except during extended storms. Earlier editions of this book joked that "off-grid power is a lifestyle, not a product," and that still applies: Understanding how your system behaves throughout the seasons is the key to reliability.

3

Examining Commercial and Industrial Solar Projects

Discover why businesses — from small shops to large industrial operations — adopt solar power, including the financial, operational, and ESG advantages that make it a strategic investment.

Find out how to evaluate commercial properties for solar power systems — whether you're installing on rooftops, parking lots, or open land — and understand the structural, visibility, and branding considerations that come with each option.

Explore the financing and ownership structures available to businesses and see how to model the financial impact, get internal buy-in, and move a project from concept to commissioning.

Chapter **10**

Seeing Why Businesses Go Solar

Solar power isn't just a residential trend; it's a major strategic tool for businesses of all sizes. Commercial solar installations now offer energy support on corporate and university campuses, in retail stores, around distribution centers, on hospitals grounds, and even near heavy industrial facilities. The business type of solar installation is commonly referred to as *C&I*, which stands for commercial and industrial. For businesses, adopting solar power is rarely an emotional or lifestyle-driven choice. It's a dollars-and-cents decision, backed by spreadsheets, forecasts, and long-term planning. In many cases, solar power becomes a predictable, low-volatility financial asset that reduces energy risk and improves a company's balance sheet.

But savings aren't the only reason companies adopt solar. Solar power can strengthen a business's sustainability and ESG profiles, help businesses meet internal climate commitments, support brand identity, and reduce exposure to rising utility rates. For some companies, solar power has become part of their competitive advantage by lowering operating costs in tight-margin industries or reinforcing a public commitment to responsible environmental stewardship.

This chapter walks you through the business logic behind making the decision to adopt solar power. You find out how companies evaluate potential savings, predict long-term energy price trends, compare financing structures, and decide whether a property is suitable for a solar power installation. And because the commercial sector is so diverse, you explore the most common use cases — from office parks to industrial plants — to help you understand the wide range of opportunities solar can unlock for businesses.

Making the Business Case for Generating Solar Energy

For businesses adopting a solar power system as an energy source is a strategic decision. Companies look at these systems the same way they look at any new equipment acquisitions, expansions, or major capital improvements — through the lens of long-term value. Being able to produce solar power can reduce operating costs, stabilize cash flow, and help companies meet public commitments on sustainability and the *Environmental, Social, and Governance* (ESG) front. When you combine these positive factors, solar power systems can become one of the most obvious and forward-thinking upgrades a business can make.

Take a look at the simplicity of the economic evaluation related to adopting solar energy production. Economic benefits can include

>> **Actual cost savings:** Commercial electricity prices tend to rise steadily over time, and many businesses operate on tight margins. Solar power generation can give them more predictable energy costs for 20 to 30 years.

>> **Higher-stakes benefits:** Although residential systems focus mostly on bill savings, commercial systems may offer better financial performance for the business overall, stronger investor confidence, and reduced exposure to future price volatility.

And when you consider the actual solar power installation, the commercial sector has the advantage of scale. Large roofs, big parking lots, and wide-open land translate into lower installation costs per watt and more energy production per system. As a result, commercial systems often have shorter payback periods than residential systems, especially when paired with tax incentives or accelerated depreciation.

Lowering operational costs

For most companies, the business case starts with a simple concept: Adding solar power production reduces the cost of electrical energy purchased from a utility, which is one of the most unavoidable expenses in commercial operations. Energy can be a major cost factor for supermarkets, manufacturers, warehouses, data centers, refrigerated storage facilities, and more. Making your own solar power helps reduce these business expenses every single month.

Table 10-1 helps illustrate what a typical commercial electric utility bill might look like before reduction by solar power. See Chapter 2 for more information about electric utility bills.

Example Commercial Monthly Electricity Costs

Cost Component	What It Includes	It's Typical Share of the Total Bill
Energy charges (kWh)	Electricity consumed over the billing period, measured in kWh	55–70%
Demand charges (kW peaks)	The highest level of power drawn at any point during the billing period, based on peak kilowatt demand	20–40%
Fixed fees	Flat monthly charges for service, billing, and maintaining the connection to the grid	5–10%

Solar power production reduces energy charges first. When paired with smart controls or storage, it can also chip away at *demand charges*: the fees utilities charge for your highest 15-minute power spike in a billing cycle.

In industries with regular demand spikes (manufacturing, HVAC-heavy buildings, or grocery refrigeration, for example), combining solar power production with battery storage can dramatically reduce demand charges, which can sometimes be more than the solar savings alone.

Lowering operational costs improves net operating income (NOI), which matters for corporate owners and for real estate investors whose building valuations depend on *NOI multiples* (a common real estate valuation method in which property value is estimated by dividing NOI by a market capitalization rate). Solar power production doesn't just cut bills; it increases the financial value of the property itself.

Locking in energy prices

Businesses also choose to incorporate solar power production because it functions as a long-term hedge against rising utility rates. Unlike residential customers, commercial users often face more complex rate structures with time-of-use

(TOU) pricing, demand charges, seasonal pricing, and sometimes *ratchets* (clauses that set the billed demand charge based on the highest peak usage in previous months instead of just the current month's usage) that can lock in high demand charges for up to a full year.

Solar power systems act like a fixed-price energy contract: Once they're installed, the sunlight that fuels the system is free and mostly predictable. A business can forecast its energy costs decades into the future with far more certainty.

This predictability has real financial value, including:

>> **Better long-term budgeting:** When you have a handle on expected energy costs, you can work a more realistic expense into your budget.

>> **Protection against fuel price strikes:** Reduced risk from sudden increases in fossil fuel prices that can raise electricity costs.

>> **Reduced exposure to policy or rate-design changes:** Less vulnerability to new utility rate structures, fees, or regulatory changes that could increase costs.

>> **More stable cash flow:** Greater predictability of monthly expenses, making it easier to plan finances and investments.

For companies with thin margins — businesses such as supermarkets, cold storage, logistics, and restaurants — stabilizing energy expenses can be the difference between profitability and loss.

The benefits you get from adding solar power production to your business isn't always cheapest in the first year after system installation. It's cheapest in later years of operation — years 5, 10, 15, and beyond. Businesses must look at total lifecycle cost reduction, not just the first month of savings.

Meeting equity, sustainability, and governance goals

Adding solar power production as a resource supports corporate sustainability strategies. Even companies without formal ESG programs often publish sustainability reports, set emissions targets, or use renewable energy as part of brand identity. A solar power system delivers highly visible, measurable progress toward these ESG goals.

Common motivations for adopting ESG strategies include:

>> **Demonstrating leadership to customers and investors:** Showing a visible commitment to sustainability and forward-thinking business practices.

>> **Reducing Scope 2 emissions:** Lowering indirect greenhouse gas emissions from purchased electricity, steam, heating, or cooling. (*Scope 2 emissions* are the emissions created at the power plant that generates the energy you buy.)

>> **Meeting internal carbon reduction or sustainability targets:** Helping the organization achieve its own climate goals, ESG commitments, or net-zero plans.

>> **Improving brand perception for environmentally conscious audiences:** Strengthening reputation with customers, partners, and employees who prioritize environmental responsibility.

>> **Qualifying for green financing or certifications programs:** Becoming eligible for sustainability-linked loans, tax incentives, rebates, or certifications such as LEED or ENERGY STAR.

>> **Supporting community and regulatory expectations:** Aligning with local climate policies, public expectations, and stakeholder pressure for cleaner energy.

Having a rooftop covered in solar panels is tangible proof of a company's commitment. Many businesses promote solar projects in marketing campaigns, investor decks, hiring pitches, and community outreach.

Large companies may adopt solar power production as part of broader energy strategies by pairing it with *electrification* (switching from fossil fuel power to clean electric power), battery storage, procurement of renewable energy credits (RECs), or even 24/7 clean energy commitments.

Exploring Common Commercial and Industrial Use Cases

Solar power production shows up in more places than most people expect, and commercial and industrial (C&I) properties are some of the biggest beneficiaries. From retail stores to warehouses to office campuses, various types of businesses use solar for different reasons — for cost savings, resilience, sustainability goals, or simply for taking advantage of large, unused roof or parking surfaces. Because most commercial buildings operate during the day (when solar power production peaks), they're often ideal solar candidates.

Across the C&I world, the solar theme is the same, and you can sum it up with this equation:

Big Roofs + Daytime Loads = Strong Solar Economics

Table 10-2 shows some common use cases for implementing solar power systems in real commercial deployments and tells why each one tends to work well.

TABLE 10-2   **Common C&I Solar Use Cases**

Use Case	Why Solar Is a Strong Fit
Retail and supermarkets	Large flat roofs, heavy daytime loads, refrigeration-driven peaks that solar helps offset
Warehouses and distribution centers	Massive roof area, minimal shading, steady energy use, motivation for ESG, some of the lowest-cost solar installations
Manufacturing facilities	High, often spiky demand; solar helps shave peaks and stabilize long-term energy costs
Office buildings and campuses	Daytime occupancy, consistent loads, strong ESG/branding value, and opportunities for solar carports
Hospitals and healthcare	Large round-the-clock energy needs; solar + storage improves resilience and reduces strain on diesel generators
Agriculture	Irrigation pumps, barns, and remote loads benefit from reduced diesel use and lower electricity costs
Parking lots and carports	Converts unused space into energy production with strong visibility and dual-use shading benefits

Optimizing form and location

Retailers, including grocery stores, big-box chains, and home improvement stores, frequently turn to solar because they have exactly the kind of space that makes installations cost-effective: huge, flat roofs with no shading. Their electricity use is highest during the daytime, when solar performs best. This is especially true for supermarkets, where refrigeration systems run constantly and contribute to high demand charges. Adding solar power production helps take the edge off those usage peaks, cutting both energy costs and operational risks.

Warehouses and distribution centers are often the "perfect storm" for commercial solar. They typically have enormous, unobstructed roof real estate, minimal rooftop equipment to work around, and energy patterns that stay fairly consistent throughout the year. This makes these installations some of the lowest-cost, highest-ROI (return on investment) project types in the entire C&I space. For many Fortune 500 companies, adding solar power production to distribution centers is the starting point for further enterprise-wide solar adoption.

Addressing operational characteristics

Manufacturing facilities often use the most electricity of any customer class, and not just a lot of it, but in sharp spikes that trigger expensive demand charges. Even if solar power production can't cover the whole load, it can significantly reduce those usage peaks, especially when paired with battery storage. For manufacturers dealing with unpredictable utility rates, solar also provides the stability of a known long-term cost.

Hospitals and healthcare facilities operate 24/7 and require uncompromising reliability. They can't depend solely on solar power production, but solar power combined with energy storage can reduce strain on diesel generators, cut operating costs, and support resilience goals. Many hospitals now use solar power production as part of a *microgrid strategy* that integrates the solar power with diesel generators and battery storage.

Agricultural operations rely heavily on electricity for irrigation pumps, cold storage, greenhouses, and equipment such as conveyors and grain dryers. Solar power production can significantly cut diesel generator use in remote areas and is increasingly used to stabilize costs for farms dealing with rising utility prices.

Putting on the ESG showcase

Corporate offices and university-style campuses benefit from solar power installations in two ways: strong daytime generation and strong branding. These types of buildings usually have steady daytime usage, plus parking lots that can be converted into solar carports. The clean-energy installation becomes a visible, tangible ESG asset for employees, investors, and visitors.

Also, solar carports turn parking lots into productive energy assets. Although more expensive than rooftop solar due to steel canopy structures, they offer dual benefits: shade for vehicles and a highly visible sustainability statement. Schools are using elevated solar structures (like car ports) specifically for the benefit of shade. If your business wants solar power production that doubles as a public ESG showcase, parking lot carports offer some of the strongest branding value per kilowatt.

Corporate sustainability, whether formal ESG reporting or informal environmental commitments, often tips the balance when strategizing for a beneficial solar installation. An evident solar power presence provides a measurable, highly visible way to reduce operational emissions, meet internal climate goals, and show customers and investors that the company is investing in long-term resilience.

Knowing What Makes a Good Solar Site Candidate

Not every commercial property is built for solar power system installation, but more qualify than many owners realize. Unlike residential installations, commercial solar decisions hinge on a blend of physical space, structural capacity, energy profile, and financial context. In the business world, "a good solar site" doesn't just mean "sunny roof." It means a property where solar power production can, in turn, produce meaningful savings, align with operational goals, and be built without triggering expensive upgrades or operational headaches.

In practice, companies look at a handful of big factors when shopping for a solar power installation location:

>> **Open space (rooftop or ground).** The condition and age of the roof, shading, local energy costs, load profiles, and how the building is accessed all apply to this factor.

- The available space determines how large a system can be and how much of the facility's energy use solar power can realistically offset.

- Roof condition matters because solar is a 25- to 35-year asset; pairing a near-end-of-life roof with brand-new panels is an expensive mistake.

- Shading, too, plays a major role: Even with modern optimizers and smart inverters, consistent shade can cut deeply into system performance and extend payback times.

>> **Characteristics of property use.** The building's energy profile is just as important. A site with high daytime usage and stable *load patterns* (in which electricity use is more evenly distributed over time) usually provides a better solar installation candidate than a site with low or irregular use.

- Businesses that use power steadily during the day — such as warehouses, retail stores, schools, and office buildings — tend to get the most beneficial financial outcomes because their loads naturally align with solar generation.

- Facilities with large evening peaks or irregular patterns may still benefit from a solar power installation, but they often need storage to smooth those spikes and capture the highest savings.

>> **A business model match.** Because commercial properties can vary so much — from warehouses to hospitals to heavy-manufacturing facilities — the criteria for a good solar power installation candidate can look different depending on the business model. For example, a hospital or data center with constant demand needs solar collection paired with energy storage, and daytime-only facility (like a school or office building) can benefit from solar collection alone.

>> **Interconnection feasibility.** The interconnection aspect can make or break a solar power project, even when the physical site is perfect. If the utility requires transformer upgrades, feeder line improvements, or months of engineering review, timelines stretch and budgets swell. Many installers identify these issues early, but they're still a factor businesses must consider before moving ahead.

Almost all commercial evaluations for solar power installation sites revolve around these foundational questions:

>> **Can a solar power installation fit the site physically?** The business determines whether the candidate site has the available area and necessary infrastructure to support the installation.

>> **Will it pencil-out financially?** The business must calculate whether a return on the investment can come within a reasonable timeframe after installation. That is, the installation must pay for itself on a timeline that works for the business.

>> **Will it integrate smoothly with ongoing operations?** The business considers the location's longevity (service to business needs over the foreseeable future) and its capacity to accommodate the solar power installation without adverse effects on access or operations.

To make these considerations more concise, Table 10-3 gives you a quick reference that summarizes the most important site factors.

A site that checks several of the boxes in the table with a positive evaluation is usually a strong candidate for a commercial solar installation. Local electricity rates shape the economics as well. Higher rates and significant demand charges increase the value of every kilowatt-hour a solar power system produces.

Key Factors That Make a Commercial Property Solar-Ready

Factor	Why It Matters	What Good Looks Like
Available space	Determines system size potential	Large flat roof, open parking lot, or unused land
Roof condition	Solar lasts 25+ years; roof must match	Roof <10 years old, structurally sound, minimal repairs needed
Shading	Directly reduces production	Little to no shading from HVAC units, trees, or nearby buildings
Load profile	Impacts payback and rate savings	Strong daytime usage; consistent year-round demand
Utility rate structure	Determines how solar saves money	High electricity rates, favorable demand-charge behavior, TOU alignment
Interconnection ease	Affects timeline and cost	Nearby transformer capacity, minimal upgrades required
Sustainability or ESG goals	Influences decision-making	Company publicly reports emissions or has internal decarbonization goals

Chapter **11**

Enhancing the Power System You Have

Adding solar to a commercial property isn't just about reducing energy bills; it's about making your existing power system stronger, more resilient, and more efficient. Businesses often have unused real estate that can be turned into productive energy infrastructure: wide-open rooftops, large parking areas, or unused land around the perimeter of the property. Each of these spaces offers advantages and requires unique design considerations, but all can support a successful commercial solar project when evaluated properly.

This chapter walks you through the main installation pathways that businesses use today — from traditional rooftop systems to striking solar carports to large ground-mounted arrays. Along the way, you find out what determines whether a property is structurally suitable, how the age of existing infrastructure and access affect cost and timelines, and why visibility matters more in the commercial world than many people expect.

Adding solar power capability isn't only an energy investment; for many companies, it's also a public commitment to operational excellence, sustainability, and modern infrastructure. Whether your business is expanding existing solar energy capacity or just exploring options, this chapter gives you the tools to match the right solar approach to your property.

Adding Solar Capability to Your Property

When a business decides to explore solar power production, the first question is usually, "Where would we even put it?" Commercial properties typically have three main options: rooftops, parking lots, and open ground on or near the facility. Each option comes with different strengths, constraints, and visibility impacts. But all three can unlock meaningful energy savings when designed well. Here are some specific considerations for solar installations:

>> **On the roof:** Many companies start with the roof as a potential solar power installation site simply because it's already there. Commercial roofs tend to be large, flat, and structurally simple, which makes them ideal solar real estate. If the roof is in good condition and has enough open space between HVAC units, vents, and skylights, rooftop solar is often the lowest-cost option.

Rooftop installations under these conditions are generally straightforward, minimally disruptive, and hidden from public view, which can be a plus for companies that want functional clean power without changing the look of their property.

>> **In the parking lot:** Parking lots offer a different value proposition. Solar carports are usually more expensive than rooftop systems because of the steel canopy structures required, but they bring unique benefits: They turn an everyday parking area into a productive energy asset; they provide shaded parking (popular in hot climates); and they act as a highly visible sustainability statement.

For many companies, solar carports serve two purposes at once: clean energy infrastructure and an unmistakable environmental, social, and governance (ESG) signal to customers, employees, and investors.

REMEMBER

>> **On the ground:** If neither rooftops nor parking lots offer enough space for a solar installation, ground-mounted arrays are the next option. For facilities with open land — such as farms, warehouses, distribution centers, data centers, and manufacturing facilities — ground mounts can support very large solar power systems.

Ground-mounted solar power systems are usually the most flexible in terms of layout and tilt angle, which can improve energy production. Businesses can also build them away from operational areas to avoid disruptions. The trade-off is land use: Not every business has spare acreage, and some prefer to reserve ground area for future expansion.

You don't need to pick just one area for solar power installations! Many commercial solar projects use a hybrid approach: rooftop panels to maximize existing space plus solar carports or a small ground-mounted system to expand capacity when the roof alone can't meet energy goals.

TIP

Rooftop solar power systems

Large commercial roofs tend to be flat or have a low slope, which gives designers enormous flexibility in how they orient and space the solar panels. Because the structure already exists, rooftop projects typically have the lowest installation cost among commercial options, making them a go-to choice for businesses focused on maximizing financial returns.

Overall, rooftop systems remain the backbone of the commercial solar market because they're cost-effective, scalable, and rarely require businesses to make trade-offs with their existing space.

Figuring the useable roof space

A usable commercial roof doesn't have to be perfect, but it does need enough open, unobstructed space to support a solar array. HVAC units, vents, skylights, parapet walls, and fire-code access lanes all reduce the footprint available for a solar installation, and commonly, only 60 to 90 percent of the roof may be usable after you figure in required clearances. Still, even a moderately sized commercial roof can host hundreds of kilowatts of solar capacity, and large warehouses routinely exceed a megawatt or more. Figure 11-1 shows an example installation that uses the ideal parts of the roof.

FIGURE 11-1: Example roof installation with room for infrastructure and shading between panels.

Oak City Drone/Adobe Stock Photos

Evaluating timing and roof conditions

Constructing rooftop installation projects is the least disruptive for a business. Most of the work happens above the building's operations, and installers can often avoid interfering with daily business activities. For companies that want solar without altering the public-facing appearance of their property, rooftop installations are the most discreet option.

These factors matter when looking at a rooftop solar installation:

>> **Roof age and condition matter as much as roof size.** Solar arrays are designed to last 25 to 35 years, so installing them on a roof that has only a few years of life left is a recipe for frustration and unnecessary cost. Removing and reinstalling panels during a re-roof is possible, but it's expensive and disruptive. For that reason, many businesses pair a planned re-roofing project in conjunction with a new solar installation so the timelines match, which avoids future interruptions.

>> **Flat roofs also make structural engineering easier.** Modern commercial racking systems rely on *ballasted mounting*, which uses weight rather than roof penetrations to hold the panels in place. This reduces the risk of leaks and simplifies installation. In high-wind regions or on lightweight structures, crews may need a mix of ballast and mechanical anchors, but even then, most commercial roofs can support solar system structures with minimal reinforcement.

>> **Shading is another key factor.** While optimizers and smart inverters help reduce the impact of partial shade, consistent shading from rooftop equipment or adjacent buildings will still lower output. A site survey and shade analysis tool (often a digital layout using aerial imagery or onsite scanning) helps determine the best zones for panels and identifies areas of the roof to avoid.

Parking lot solar installations

When a business wants a solar power system but doesn't have enough usable roof space — or wants to make a visible statement about sustainability — parking lot solar canopies, often called *solar carports*, become an attractive option. Unlike rooftop systems, carports can turn single-use space into dual-purpose infrastructure: They generate electricity while shading vehicles, reducing heat buildup inside cars, and improving the customer or employee parking experience. In hot climates, shaded parking alone can be a major selling point for adding a solar carport. Figure 11-2 shows an example of a parking lot installation.

FIGURE 11-2: Example carport installation.

Henk Vrieselaar/Adobe Stock Photos

Consider these specifics about solar carport systems; they:

>> **Cost more than rooftop systems.** The steel canopy structures add material, engineering, and labor expenses, and the foundations or footings must be carefully designed to handle wind and snow loads. But for many businesses, the value goes beyond the simple financial payback.

>> **Are highly visible, making them one of the strongest branding tools in commercial solar.** A row of solar canopies over employee or customer parking sends a clear, public message: This company invests in clean energy and modern infrastructure.

>> **Offer design flexibility that rooftops sometimes lack.** Tilt angle and orientation can be optimized for energy production, and the spacing between rows can be adjusted to accommodate traffic flow, fire lanes, delivery zones, or ADA accessibility.

>> **Present an opportunity for multiple ESG objectives.** Many businesses also incorporate EV charging stations directly into their carport installations, allowing one project to support multiple sustainability goals at once.

The construction process is more noticeable than rooftop work but still manageable. Installers must temporarily close parking sections as the structures go up, and businesses usually phase the work to minimize disruption. After installation, solar carports require little maintenance: They're engineered to withstand decades of weather and heavy use.

For businesses that want functional solar with strong ESG visibility, parking lot systems offer a compelling combination of utility and public-facing impact. They're not the cheapest option, but they often deliver the greatest value in terms of branding, customer experience, and multiuse infrastructure.

Ground-mounted solar arrays

When rooftop space is limited or parking lots are already spoken for, ground-mounted solar arrays offer flexibility and often the greatest capacity. Many commercial and industrial properties — such as warehouses, distribution centers, data centers, farms, or manufacturing sites — have underused land nearby that businesses can turn into a large-scale solar resource. Ground mounts allow designers to choose the ideal orientation, tilt angle, row spacing, and layout, which often results in higher production per panel compared to rooftop or carport systems.

For businesses with the right land profile, ground mounts can be the most powerful and customizable solar option. They maximize production, simplify expansion, and provide the kind of design freedom that rooftops and carports simply can't offer.

A big advantage of ground mounts is scalability. With open land, businesses can build systems sized precisely to their energy needs, even expanding them later as operations grow. Figure 11-3 depicts a ground-mounted solar power system. Ground mounts also simplify maintenance: Without roof obstructions or canopy structures, crews can easily access wiring, inverters, and racking, and manage surrounding vegetation with routine mowing or low-maintenance ground cover.

FIGURE 11-3:
Example ground-mount installation.

j_nameera_h/Adobe Stock Photos

Of course, lack of appropriate land area is the biggest barrier to potential ground-mounted solar power. For example,

>> Many businesses simply don't have spare acreage, or the land they do have is earmarked for future expansion.

>> Where land is available, soil conditions and grading matter. Rocky soil, steep slopes, wetlands, or sites with poor drainage can increase construction costs.

>> In some regions, fencing or wildlife considerations may exist, and businesses should factor in the aesthetic impact of a visible ground array.

Ground mounts come with the opportunity to integrate *solar trackers*, which are structures that rotate the panels to follow the sun throughout the day. Trackers increase energy production but add cost and mechanical complexity, so they're more common in utility-scale and large industrial settings than in smaller commercial projects.

Because ground-mounted arrays can be separate from the main business facility, installation is often less disruptive. Crews can stage equipment and work independently of daily operations, and they can trench electrical connections back to the building outside of peak business hours.

Knowing What Factors to Consider

Before a business commits to any type of solar power installation, it needs to understand the practical, structural, and operational factors that determine whether the project will succeed. Unlike residential systems, where roof condition is usually the only major constraint, commercial properties involve more variables. Large buildings have heavier mechanical equipment, more complex electrical systems, and long-term operational needs that must be protected throughout the life of the solar array. The goal isn't just to "fit solar on the property," but to integrate it in a way that supports reliability, safety, and long-term savings.

When determining a go-ahead for a solar installation, most businesses begin by assessing the following aspects of existing infrastructure and business operations:

>> **The structural integrity of the roof or the supporting structures for carports.** Commercial roofs are designed to carry specific loads, and solar adds both weight and wind uplift considerations. Before any installation, a structural engineer evaluates whether the existing roof can support the added load or whether reinforcements are needed. For ground mounts,

soil conditions and grading determine the type of foundation that a solar system installation will require.

>> **Roof weight limits and age play a central role as well.** A roof nearing the end of its useful life isn't a good solar candidate unless the business is prepared to re-roof before installation. Because commercial solar systems last 25 to 35 years, the roof underneath them should last just as long. Similarly, reinforcing older structures may add to project cost, but it can also extend the building's life and ensure the system is safe and code-compliant.

>> **Ready access is another major factor.** Installers need room to move equipment, stage materials, and plan construction in a way that doesn't interfere with daily business operations. On rooftop installations, access includes ensuring clear walkways for maintenance and keeping required fire lanes unobstructed. For parking lots and ground mounts, maintaining traffic flow and safe working areas may require phasing the project. Businesses that rely on constant customer or delivery activity need a plan that minimizes disruption.

Planning for a commercial solar installation also introduces a set of considerations that go beyond engineering: visibility and branding opportunities. Here are the trade-offs:

>> Rooftop systems are discreet, but carports and ground mounts are often highly visible.

>> Many companies treat solar as an extension of their brand identity, something employees and customers will see daily. For businesses with ESG goals or sustainability initiatives, the placement and design of the installation can reinforce company values and public commitments.

>> Carports, in particular, double as shade structures, charging-station hubs, and visual proof of a company's investment in clean power.

What If You Don't Own the Building?

Many businesses — especially retail stores, offices, warehouses, and industrial spaces — don't own the properties they operate from. That doesn't automatically rule out adding a solar power installation, but it does change the process. When you're a tenant, the biggest question isn't "Can a solar power system save us money?" but "Do we have the right to install a solar power system?"

Solar power systems alter the roof structure, electrical system, and sometimes the property's long-term maintenance needs, so landlords need to approve any installation. In many cases, landlords may need to be co-signers on permits or utility applications. Long-term leases, predictable operations, and a cooperative property owner make tenant-driven solar power installations much more feasible.

Businesses that lease typically go one of three routes:

>> Some tenants install and own the system themselves under a roof lease or solar addendum.

>> Other tenants work with the landlord, who installs and owns the system, and may pass savings along through reduced electric rates or common-area maintenance charges.

>> In multi-tenant or industrial parks, a third party sometimes owns the solar power system and sells power behind the meter to one or more tenants.

The key to successful solar power installation as a tenant is to start the conversation early. If you don't control the roof or electrical infrastructure, your landlord's support is essential. A quick review of your lease, especially sections related to structural changes, roof access, and utility connections, will tell you how realistic your solar plans are.

Structural Integrity and Practical Constraints

Before a commercial solar project can move forward, the property itself has to be ready: structurally, logistically, and operationally. This step isn't as flashy as talking about energy savings or ESG goals, but it's one of the most important parts of the planning process. A building or site that can support a solar power installation physically and safely is far more valuable than one with perfect sunlight but major construction hurdles.

Solar power systems add weight, wind uplift forces, and electrical equipment to a property, and installers must confirm that the existing structures — which solar systems are built on and around — can handle all these additions. Verifying that properties are ready for a solar power installation involves evaluating the following:

>> **Infrastructure viability and lifespan:** On commercial buildings, a structural engineer reviews the roof framing, decking, age, and condition to verify that solar loads won't compromise the building. Older roofs may need reinforcement or replacement before solar panels go on. Even roofs that look perfectly fine from the outside may be nearing the end of their lifespan, so many companies time a solar power system installation to coincide with reroofing to avoid refitting a system later.

>> **Uninhibited space:** Structural considerations go beyond just whether the roof is strong enough. The layout of mechanical equipment, parapets, skylights, and drainage paths all influence how panels can be installed. A roof covered with HVAC units, vents, and odd angles still may qualify for a solar installation, but the usable space and layout will dictate system size and cost.

>> **Access characteristics:** Commercial solar installations require moving thousands of pounds of equipment, panels, racking, ballast, conduit, and inverters onto the property. Installers need safe pathways for crews, staging space for materials, and crane or lift access for rooftop systems.

If equipment can't be craned in because of overhead lines, narrow streets, or landscaping obstacles, installers may need alternative lifts or to break down equipment into smaller loads. That doesn't make a project impossible, but it can affect the budget and timeline.

>> **Fire safety and code requirements:** Most jurisdictions require designated fire access pathways on roofs, spacing between panel blocks, and clearances around mechanical equipment. These requirements aren't optional; they're life-safety rules that installers must integrate into the system layout, which is why solar installations rarely cover 100 percent of a roof, even when it appears to have plenty of space.

When installers talk about site readiness, they're looking at all these issues at once: roof age, weight capacity, layout, access, fire pathways, and any structural upgrades that might be needed. The goal isn't to find reasons to reject a site, it's to understand how large a solar power system the property can host without compromising safety or driving up costs. Many buildings with mild limitations still end up hosting excellent solar projects after the design is tailored to their layout.

Thinking about Visibility and Branding Opportunities

For many commercial properties, the value of a solar power installation goes beyond energy savings. A solar installation can become a visible statement of a company's brand, values, and long-term strategy. In a world in which customers, investors, and employees increasingly care about sustainability, the decision to "go solar" can quietly (or not so quietly) boost your public image.

Advertising the decision to add solar power

Some businesses treat solar the same way they treat upgraded lobbies, signage, or architectural features: as an investment in how the property communicates who they are. A rooftop solar panel system isn't always visible from the street, but drones, marketing materials, and website photography make up for that quickly. And when the solar power system is placed in parking lots or ground-mounted arrays, customers can see it every time they visit the business site. For retail, hospitality, education, and healthcare facilities, this visibility can reinforce trust and alignment with stakeholder values.

Branding opportunities from solar power installations come in different forms:

>> Parking-lot solar structures often present space for tasteful signage or messaging.

>> Corporate campuses sometimes add educational displays in lobbies that show real-time solar production, carbon reductions, and system performance, turning clean energy into a talking point for tours, community events, and investor meetings.

>> Even ground-mounted systems can coexist with thoughtful landscaping, walkways, and interpretive signs that tell the story of how the business is reducing its environmental footprint.

Identifying the business with green initiatives

Solar power installations also support internal brand identity. Many companies use their installations as part of employee engagement efforts, sustainability task forces, green teams, or volunteer events tied to Earth Day. The incorporation of solar power becomes visible proof of a company's commitment to long-term responsibility and can help with employee recruitment and retention — especially for younger job-seekers who increasingly evaluate employers on environmental performance.

The key to using a solar power installation as a branding opportunity is intentionality. A company that treats solar purely as a back-of-house infrastructure upgrade may miss out on its storytelling power. A company that thinks strategically about visibility — even if it's only through photographs, case studies, or internal communications — can get far more value from the same installation. Solar produces power, but it also produces a narrative: one that signals stability, foresight, and alignment with modern expectations for corporate responsibility.

Chapter **12**

Choosing Financing and Ownership Options

Commercial solar power doesn't begin on the roof, it begins on the balance sheet. Before businesses order any panels or start the engineering process, they must decide how they want to pay for a solar power project, who will own the system, and how the financial benefits can flow over time. This chapter walks through the financing choices that shape both the economics and strategy behind commercial solar projects.

You find out what happens after businesses make the decision to pursue solar power, including how to model the financial impact of adding solar, build internal support (from finance teams to executives), and manage the installation process after a contract is signed. You get a clear sense of not only how businesses finance solar, but also how they move from a promising proposal to a fully commissioned, grid-connected system.

Paying for Solar versus Leasing

For commercial solar buyers, the very first fork in the road is deciding whether your business will own the solar system or lease it from someone else. Both paths deliver renewable energy and long-term savings, but they differ dramatically in tax treatment, cash flow, and responsibility for the equipment. In fact, this ownership choice often matters more to a company's bottom line than the specific panels, inverters, or mounting system chosen later.

Unlike homeowners, businesses have a broader menu of funding options and incentives. They can capture tax credits directly, lease systems from third parties, sign long-term power purchase agreements (PPAs), participate in community solar markets, or bundle solar with other energy-efficiency upgrades under specialized financing programs. Each option affects the company's cash flow, tax position, and operational responsibilities differently, which is why choosing the right one is often more impactful than choosing the panel brand.

Considering outright ownership

When a business purchases a solar system, whether paid in cash or financed, it also owns every financial benefit that comes with it. That includes the federal Investment Tax Credit (ITC), accelerated depreciation, reduced electric bills, and any applicable state incentives. Ownership offers the strongest long-term return, which is why companies with healthy tax appetites often prefer it. But it also requires capital, balance-sheet capacity, and a willingness to own and maintain a long-term asset.

Ownership gives companies full authority over system design, branding opportunities, metering, and the ability to expand or modify the system later.

If your business can use the tax benefits, ownership is almost always the financial winner. If you can't use them, leasing structures allow someone else to monetize those credits and pass part of the value back to you as lower energy pricing.

Looking at leasing

When a business leases a solar power system or signs a long-term power purchase agreement (PPA), a third party owns the equipment and takes the tax benefits that come with it. The customer typically pays nothing upfront, enjoys predictable energy pricing for years, and avoids responsibility for system maintenance and performance. This leasing route can make solar power installations feasible for organizations that want to avoid tax liability or that have no desire to own energy infrastructure.

The trade-off is that the total financial value of using a solar power system is usually lower than with ownership because a portion of the benefits goes to the system owner. Also, lease and PPA agreements usually limit operational flexibility and require landlord-like approval from the system owner. Neither type of agreement is inherently better, but they simply fit with the needs of different organizations.

Examining the differences

In practice, many businesses approach the owning versus leasing decision for a solar power system the same way they approach buying versus leasing a fleet vehicle. If you want the lowest long-term cost and are comfortable owning the asset, buying is usually the better deal. If you want simplicity, no upfront cost, and predictable budgeting, leasing or signing a PPA offers a clean, low-risk option.

To help you see the differences between owning and leasing a solar power system at a glance, Table 12-1 offers a clean comparison of the two most common commercial solar models.

TABLE 12-1 **Comparing Ownership and Leasing Models for Commercial Solar**

Feature/Consideration	Ownership Model	Leasing/PPA Model
Upfront cost	Higher	Typically low, can even be $0
Tax credit benefit	Business	Third-party owner
Electricity price	Free after system pays back; lowest long-term cost	Fixed or escalating rate per kWh over contract term
Maintenance responsibility	Business (self or outsourced)	Third-party owner
Balance-sheet treatment	Asset and liability (if financed)	Off-balance-sheet for many leases/PPAs
Best for	Companies with tax appetite and long-term capital strategy	Companies without tax appetite or upfront capital, nonprofits, or public entities
Long-term savings	Highest	Moderate; traded for convenience and no upfront cost

Using Tax Incentives for Business

For many companies, tax incentives are the single biggest reason that commercial solar projects make financial sense. Unlike rebates, which reduce upfront cost, tax incentives reduce the taxes your business owes, sometimes dramatically. These incentives can shorten payback periods, improve cash flow, and turn projects that looked marginal on paper into strong performers. Understanding how tax incentives work helps you decide not just whether to go forward with solar power, but how to structure ownership so that your business gets the most value.

Two major tax incentives at the federal level for businesses are

>> **The federal Investment Tax Credit (ITC).** This is the backbone of commercial solar incentives for businesses in the United States. As of this writing, the ITC allows companies that invest in solar power systems to claim a percentage of the system's cost as a credit against federal income taxes. The exact percentage in any given year depends on current federal law and whether the project meets certain requirements (such as using domestic content or being located in an energy community).

What's important to note is that the ITC operates as a dollar-for-dollar reduction of taxes owed, not a deduction that merely lowers taxable income. If your company builds and owns the system, the credit belongs to you.

>> **The Modified Accelerated Cost Recovery System (MACRS).** This is the most notable system by which commercial solar projects benefit from accelerated depreciation. MACRS allows businesses to recover most of the solar power system's value over a short period, improving near-term tax benefits. In many cases, bonus depreciation has also been available, which lets companies depreciate a significant portion of the system cost in the first year.

These depreciation rules shift occasionally, so check the current year's IRS guidance or speak with a tax professional to understand how much accelerated depreciation your business can claim.

Some companies, however, can't use tax credits or depreciation directly, perhaps because they're nonprofits, they operate at low margins, or they simply don't have enough taxable income. In those cases, tax incentives become a major factor in deciding between ownership and a third-party model like a lease or PPA. When a third party owns the system, they take the tax incentives and build those benefits into the price of the energy they sell you. You don't get the credit directly, but you still benefit indirectly through lower electricity prices.

State and local incentives can also play a significant role. Some states offer performance-based incentives paid out over time, while others offer one-time grants, rebates, or even property-tax exemptions for renewable-energy systems. Utility programs — where they exist — may provide additional payments or credits for solar power generation. Because these incentives vary widely by region, and change frequently, most businesses work with their engineering, procurement, and construction (EPC) contractor or financier to build a complete set of incentives tailored to their location.

Making a Solar Option Happen

After you understand the basic financing options and tax incentives, the next step is turning "this could work" into "we're doing this." For a business, that doesn't happen just because someone likes solar power in principle. It happens when the numbers are clear, the risks are understood, and the decision-makers see how the project fits into the company's larger financial and operational plans. Adding solar is a cross-functional decision that requires stakeholders to exist across finance, operations, marketing, and other teams.

Most commercial solar projects move through a similar pattern:

» A preliminary proposal sparks interest, and someone inside the company digs into the numbers.

» Related questions start coming in from finance and leadership in various departments, and the investigation continues.

» Eventually, a champion emerges inside the organization and pushes the project across the finish line.

This section walks you through the parts of the decision-making process that you can control: how to model the impact of adding a solar power system for your business and how to build the internal support you need to get a project approved.

Modeling the impact of adding solar to your business

Modeling a solar project is about answering a few simple questions, including

» How much will adding solar power to the business cost?

» How much will having supplemental solar power save the business?

» What period of time elapses before the savings make up for the costs?

The tools and spreadsheets you use to figure out the answers to these questions can get fancy, but the basic structure for building a model of the impact of adding solar energy production stays the same no matter how large your facility is.

Most companies don't build models from scratch. Instead, they review and stress-test the models provided by installers, financiers, or internal energy teams. You don't need to become a project finance expert, but you should be comfortable asking questions about how the proposed performance of a solar power system is derived. What utility rate escalation did you assume? How conservative is the production estimate? How did you treat maintenance costs over time?

You analyze the impact of adding solar power to your business by following these general steps:

1. **Establish a baseline for existing energy consumption.**

   Energy usage data includes your current electricity usage, your rate structure, and your annual expenditure.

2. **Figure the costs and savings associated with the proposed solar power system.**

   The layers of this analysis include expected annual production, the portion of your energy usage that solar production will offset, and how your utility bill changes under your specific rate plan (for example, time-of-use pricing, demand charges, or net-metering rules).

   You also want your impact model to look at

   - *Installed costs:* broken out into equipment, labor, and soft costs

   - *Incentives:* tax credits, depreciation, and any state or utility programs

   - *Resulting net cost:* after incentives

   - *Annual electricity savings:* based on how much grid energy the solar offsets and what you would have paid without it

   - *Payback period:* how many years until cumulative savings equal your net cost and common metrics like net present value (NPV) or internal rate of return (IRR)

   A good installer or EPC contractor can provide a pro forma document showing year-by-year savings, but you should still understand what's under the hood of the system that offers these savings.

3. **Change up your modeling to account for various operational scenarios.**

   Good models also look at various scenarios and not just a single forecast. You might compare a conservative case (lower production, modest utility rate increases) with an expected case (average production based on typical weather and planned utility rate changes) and an upside case (higher-than-expected production combined with faster utility rate increases or additional incentives).

The point isn't for the model to predict the future perfectly; it's to understand how resilient the project can be if conditions change. A project that still looks reasonable under conservative assumptions is much easier to defend internally than one that only works if everything goes perfectly.

If you're a tenant or don't control the roof, your modeling may focus less on owning an asset and more on comparing contract options: different PPA rates, *escalators* (annual price increases built into the contract), or other terms. In that case, you treat adding solar power like a long-term energy supply contract and evaluate it next to your expected utility costs over the same period.

Ask your installer or financier for both a *simple payback* (years to break even) and at least one discounted cash-flow metric such as *net present value* (NPV) or *internal rate of return* (IRR). Finance teams are used to seeing those numbers and will take the project more seriously if it's framed like any other capital investment.

Getting buy-in from your organization

Even the best solar power installation proposal can stall if it doesn't have a clear internal champion. In most organizations, that champion is someone who understands both the technical side of solar power system and the political side of making a big business decision. If you're the champion, your job isn't just to like the idea; it's to help others see why it fits the business.

Different stakeholders in your business care about different things. Finance leaders want to know how a solar power installation affects cash flow, return on capital, and risk. Operations and facilities teams care about the impact on infrastructure, maintenance, and reliability. Sustainability or ESG leads care about reduced carbon emissions and how the project supports public commitments. Executives may be thinking about brand identity, customer perception, and long-term energy strategy. A strong solar project pitch doesn't drown everyone in the same details; it gives each group what they need to say "yes."

An effective tool for supplying the needed details is a simple one- or two-page summary that translates the project model into the plain language that speaks to various stakeholders. That summary highlights

» What type of solar installation you're proposing to build (size, location, and ownership model)

» Rough project cost and financing plans

> » Annual and lifetime savings, including incentives

> » Key risks and how they're being managed (roof condition, performance guarantees and maintenance responsibilities)

> » Any additional benefits of adding solar power, such as ESG impact, visible branding, or employee comfort (for example, in the case of carport solar installations that provide shaded parking)

A solar project proposal becomes stronger with each round of questions you answer. Instead of trying to answer every question in a single meeting, think of creating buy-in as a series of small conversations. You might walk through the pro forma figures with finance, invite facility representatives to speak directly with the installer about roof loading and access, and give executive leadership a higher-level overview focused on strategy and optics.

When solar project modeling and buy-in come together, a solar project stops being an abstract sustainability idea and becomes a concrete business decision. That's the point where you're ready to move into the next phase: actually building your system, selecting partners, and moving from proposal to construction.

REMEMBER

Building Your System

After your business decides to move forward with implementing a solar power solution, the next phase is turning a well-modeled, internally approved project into a real installation. This part of the process looks less like financial analysis and more like a real-life construction project because that's exactly what it is. You work with technical partners, review designs, coordinate schedules, and make decisions that affect both the roof (or other installation site) and your business's daily operations.

Setting the groundwork for a solar project

The good news is that you don't need to become a solar engineer to manage the construction phase. Specialized partners who live and breathe permitting, engineering, procurement, and construction help you with the details of design and installation. Your role is to choose the right team, understand the basic steps, and keep the project aligned with your business's operational needs. The clearer you make upfront details — about how your facility runs, where equipment can go, and what constraints your operations have — the smoother the project construction will be.

Solar projects tend to follow a predictable pattern that involves

>> Selecting your EPC or contractor

>> Finalizing the design

>> Securing permits and utility approvals

>> Scheduling construction around your business operations

>> Commissioning the system so it can safely deliver power

Some parts of this pattern require more involvement from you than others, but they all benefit from good communication between project leaders, sponsors, and operatives. The more transparent you are with your EPC about your roof condition, electrical infrastructure, and operational hours, the fewer surprises appear later.

A little preparation at the project setup phase goes a long way to ensure the success of a solar project installation. Understanding the roadmap for getting through construction helps ensure your project meets its schedule, stays within budget, and satisfies the quality standards your business expects.

Hiring an EPC

For commercial solar projects, the most important partner you choose is the EPC: the engineering, procurement, and construction firm responsible for designing your system and building it safely. Picking the right EPC can make your project run smoothly and be predictable; choosing the wrong one can introduce delays, unexpected costs, and operational headaches.

A strong EPC does far more than bolt panels to a roof. They evaluate your site, model expected production, prepare engineering drawings approved by a licensed engineer, navigate permits and utility requirements, procure all equipment, oversee construction crews, and ultimately deliver a system that meets building and installation code requirements and performs as promised.

Most businesses evaluate EPCs using a blend of experience, technical capability, safety record, financial stability, and communication style. When you choose an EPC partner, look for these characteristics and practices:

>> **Experience with similar solar installations:** Years in business and total megawatts installed are helpful reference points, but what matters most is whether they have experience with projects similar to yours: rooftop versus ground mount, ballasted versus penetrating racking, carports, manufacturing

facilities, logistics hubs, cold storage, or multimeter campuses. An EPC who has built dozens of systems for facilities like yours will anticipate issues you haven't even thought to ask about.

>> **Careful upfront inspections:** A good EPC will walk your site early and look closely at details such as roof condition, electrical service size, structural loading limits, fire code pathways, and interconnection options. This early diligence shapes everything that happens later, from whether the system needs structural reinforcement to whether a service upgrade is required. Don't be alarmed if a strong EPC asks challenging questions or identifies constraints you didn't know you had. That's their job, and good EPCs surface these realities before construction, not during it.

>> **Project management discipline:** Solar projects involve coordination across engineering, scheduling, procurement, permitting offices, utilities, and on-site construction crews. You want an EPC who communicates clearly, responds quickly, and provides dependable timelines. Ask how they handle delays (because something *will* come up), how often they provide progress updates, and whether you'll have a single point of contact.

>> **Good cultural fit:** The EPC will be interacting with your facilities team, your roofers, your electricians, and sometimes your operations staff. An EPC who listens well, explains things clearly, and respects how your business operates will make the project far easier to manage.

Don't hesitate to ask for references from customers with similar buildings or industries. A ten-minute call with another facility manager can tell you more about an EPC's reliability than a polished proposal ever will.

Comparing quotes from prospective partners

When you start collecting proposals, you'll quickly notice something: Proposals for two solar power systems that have the same overall engineering can look very different once you dig into the details. Making a good comparison among proposals involves more than picking the lowest bid; you need to understand what each EPC is proposing and how their assumptions, equipment choices, and scope differ. The clearer those differences are, the easier it becomes to choose the partner who will deliver the performance and reliability your business expects.

Here are some important parts of evaluating proposals and comparing quotes:

>> **Review each EPC's expected annual energy production because** this output determines your savings. A proposal offering a slightly lower price

but significantly lower annual kWh of energy output may actually cost you more in the long run.

If energy production numbers vary significantly across proposals, ask why: Shading assumptions, layout choices, and tilt angles of the solar panels' installation can create real differences in the system's performance.

>> **Look closely at equipment specifications,** such as panel efficiency, inverter type, warranties, and degradation rates. These specs shape how the system performs over 25 years.

>> **Pay attention to the assumptions behind each financial model.** EPCs often use different utility rate escalations, derate factors, or operating hours, which can make one project look better on paper without reflecting reality. Aligning assumptions helps you compare designs rather than optimistic modeling. Check these specific points across all proposals:

- *Scope clarity:* Look for whether costs of roof repairs, structural reinforcement, electrical upgrades, or access equipment (like cranes) are included or excluded.

  If a proposal seems unusually inexpensive, it may simply be incomplete. Ask for a scope clarification sheet to avoid surprises later.

- *Construction impact:* Some EPCs plan work that minimizes business disruption; others assume full access during business hours.

- *Cost structure:* The best quotes clearly break out equipment, labor, engineering, and soft costs (such as permitting fees, inspection costs, and utility interconnection applications and fees) so you can understand where pricing differences come from.

Evaluating a project timeline

Solar projects move through several predictable phases, and understanding the extent of the timeline involved helps you plan around permitting, construction, and utility approvals. Although the exact timing varies by location and system size, most commercial projects follow a similar arc from design to *commissioning* (the final functional testing and setup for operation; see the next section). Table 12-2 offers a quick reference for what happens when and who's responsible for each step.

The longest parts of a commercial solar project nearly always involve permitting and utility interconnection. Construction itself is usually the fastest phase.

Phase	What Happens	Typical Duration
Final engineering and design	Detailed structural and electrical plans, stamped drawings, fire pathway layouts, equipment selection	2–6 weeks
Permitting	Building, electrical, and fire department permit submissions and approvals	2–12 weeks
Utility interconnection review	Utility studies, paperwork, metering review, approval to construct	4–16 weeks (can overlap with permitting)
Procurement	Ordering panels, inverters, racking, switchgear, and long-lead items	2–10 weeks
Site preparation	Staging, safety prep, structural anchoring, equipment delivery	1–3 weeks
Construction	Installing racking, wiring, panels, inverters; tying into electrical system	2–8 weeks
Inspections	Authority having jurisdiction (AHJ) inspects project for code compliance and safety	1–3 weeks
Commissioning and PTO	System testing, monitoring setup, utility Permission to Operate (PTO)	1–4 weeks

Commissioning your installed solar power system

Commissioning is the final checkpoint before your system officially goes live. After construction, your EPC runs a series of electrical and performance tests that verify wiring, inverter settings, grounding, racking connections, and monitoring equipment. This final check ensures every component works as a unified, safe system.

Next comes an inspection by the *Authority Having Jurisdiction* (AHJ, the organization, office, or official responsible for enforcing codes, standards, and regulations). This inspection confirms that the project meets local building and electrical codes. After that, the utility your system will connect to performs its own review and, once satisfied, issues *Permission to Operate* (PTO, your official go-ahead to produce solar energy). Until PTO is granted, your system can't send power to the grid, even if everything is physically complete and operational.

When PTO arrives, you can switch on the system, begin monitoring energy production data, and provide basic training for your team responsible for watching dashboards and adjusting controls. The EPC typically provides a final commissioning report and remains available during the warranty period.

If production looks off in your monitoring portal during the first weeks, alert your EPC early, it's easier to resolve issues immediately after commissioning.

The work is behind you, and now you can start to enjoy your commercial solar system!

4

Utility-Scale Solar: Powering the Grid

See what makes a solar project *utility-scale,* where these projects are built, who owns them, and why they play such a major role in powering cities, industries, and entire regions.

Follow a utility-scale solar project from a blank plot of land to a producing power plant by tracking the developers, approvals, financing, and construction steps along the way.

Discover who buys all that utility-scale solar power, how they buy it, and how policies and grid conditions affect project value.

Chapter **13**

Describing Solar Power at Utility-Scale

tility-scale solar projects represent the biggest, boldest version of renew-able power. Instead of powering a single home or business with a rooftop system, these projects generate electricity at a scale that's large enough to supply thousands (or sometimes millions) of homes. When most people picture a solar farm, they imagine utility-scale solar installations in which rows and rows of panels stretch across open landscapes and feed directly into the electric grid.

Why should utilities build solar power systems this big? The grid needs enormous amounts of clean electricity, and utility-scale projects deliver it efficiently. The cost per watt of power generated by these large-scale systems can be much lower than for smaller installations. Thanks to economies of scale, the energy output is massive, and the projects support regional reliability by producing power where the grid needs it most. The number of utility-scale solar power systems has grown rapidly over the last decade, and they are now among the leading sources of new power generation worldwide.

In this chapter, you discover the benefits of utility-scale solar projects and find out the new challenges they bring about, including coordinating with utilities, securing access to transmission lines, navigating land use and environmental concerns, and working through complex financing structures. But when all these aspects come together, the impact on clean energy production is huge: A single project can offset millions of tons of carbon dioxide (CO_2, greenhouse gas) released into the air across its lifetime. This offset plays a vital role in shifting the entire power system toward producing cleaner energy.

Defining Utility-Scale

Utility-scale solar represents the largest category of solar power projects; they're designed to serve the electric grid, not an individual building or campus. The electricity these projects generate is sold wholesale, typically under long-term contracts, and delivered into transmission lines that feed cities, towns, and entire regions. These projects are the solar farms most people imagine when they think of a sea of solar panels in the desert.

Table 13-1 gives you an overview of the types of solar projects and the scope of their power production.

TABLE 13-1   **Typical Solar Project Categories and Characteristics**

Category	Typical Capacity Range	Common Settings	Powering
Residential	3–15 kW	Homes, apartments, cabins	One household or building
Commercial	15 kW–5 MW	Warehouses, schools, offices	A single business or campus
Utility-scale	5-500+ MW	Large solar farms, usually in deserts or farmland	Entire towns or regions

Where rooftop and commercial solar power systems reduce a specific customer's bill, utility-scale solar lowers the cost of power for everyone. Economies of scale allow developers to build utility-scale systems more efficiently because they can

>> Make use of currently under-used land

>> Purchase solar components and other infrastructure in bulk

>> Employ advanced tracking systems that follow the sun throughout the day to squeeze out maximum solar collection and energy production

Here's one helpful way to picture the scale of utility-scale: a 200-MW solar farm can power tens of thousands of homes. The largest projects today exceed 3 gigawatts (GW), which is enough to power a major city from just one location.

The following points are typical characteristics of utility-scale solar power systems:

>> **They require space, but actually less space than most power production sources.** A common rule of thumb is 5 to 8 acres per megawatt, including access roads and equipment areas. Developers can pair solar power production with ongoing land uses like sheep grazing or planting pollinator-friendly ground cover to support local ecosystems.

>> **They deliver power into high-voltage transmission lines so that it can travel long distances.** Proximity to suitable grid infrastructure, and the ability to obtain an interconnection agreement (covered more in Chapter 14), often determines whether a utility-scale project is even possible.

Although rooftop solar systems will always play an important role in distributed power, utility-scale solar does the heavy lifting in global decarbonization. It provides huge volumes of cleaner energy at some of the lowest costs ever seen in the power sector.

Acknowledging the Massive Impact of Utility-Scale Solar

Utility-scale solar projects represent one of the biggest success stories in modern energy production. Over just the last decade, utility-scale systems have transformed from a niche technology into one of the leading sources of new electricity worldwide. And they bring benefits on multiple fronts: lowering costs for consumers, reducing harmful emissions, and strengthening the reliability of the power system.

On the economic side, utility-scale projects make solar one of the cheapest forms of new power generation in history. Large projects spread equipment and labor costs across millions of watts of capacity, making the cost per kilowatt dramatically lower than rooftop systems. That means utilities and grid operators can purchase clean power at prices competitive with, or even lower than, fossil fuels, even before incentives are applied. These projects also bring regional benefits:

>> **Jobs** in system development, construction, operations, and maintenance support thousands of workers.

>> **Local revenue** increases because of land leases and taxes collected.

Environmentally, the impact of large solar projects is equally significant:

>> A single large solar farm can avoid millions of tons of CO_2 emissions over its lifetime. Unlike fossil plants, solar projects produce zero air pollution, no ash waste, and use very little water, which is a major advantage in arid regions.

>> Some sites also promote dual land use, such as grazing or planting pollinator-friendly ground cover beneath and between panel rows, helping restore ecosystems while generating power.

A lesser-known benefit involves grid stability. Because utility-scale solar projects connect directly to high-voltage transmission lines, they deliver energy where it's most needed and help reduce stress on older plants and infrastructure. Pairing solar power production with storage capacity — which you can explore in Chapter 17 — also makes the grid more resilient during extreme weather or power demand spikes.

Recognizing Ownership Models

Behind every utility-scale solar project lies a simple question: Who owns this thing? The answer matters because ownership shapes how a project is financed, who takes on risk, and how long the asset is likely to remain with that owner. Most large solar farms end up in one of three buckets: projects that are ultimately owned by utilities, those owned by independent power producers (IPPs), and those owned by institutional capital and infrastructure funds that treat solar like a long-term investment.

You may never see ownership shifts from the outside. A solar farm that starts its life under one owner may be sold once, twice, or even more over the course of its operating life. But understanding the basic models presented in this section helps you make sense of who is behind the projects powering the grid.

The same solar power project can move through more than one ownership model over its lifetime. An independent developer may build the installation, a utility may buy the power, and an infrastructure fund may eventually own the asset. In any case, all the electrons look identical on the grid.

Utility ownership

When people hear *utility-scale* in relationship to a solar power project, they often assume that the utility owns it. Sometimes that's true, but it doesn't usually mean the utility's own crews are out there pounding posts and pulling wire. In a utility-owned model, the utility is the end owner of the asset, even if a separate developer and engineering, procurement, and construction (EPC) team handled development and construction.

Utilities pursue ownership for a few reasons. The project becomes part of their regulated asset base; they can plan it as part of their long-term resource mix; and they earn an approved return on the investment through customer rates. From a grid-operations standpoint, utility ownership can simplify planning and dispatching the generated power because the same organization is responsible for both the power plant and the wires that carry its output.

In practice, utilities often run a competitive process — in which multiple vendors submit bids, proposals, or contract applications — to select a developer or EPC that will develop and build the project. Then, the builder either sells the project to the utility at completion or uses a build-transfer agreement in which the utility takes ownership at a specific milestone — for example, when the project is ready to operate. From the outside, the project can look like a "utility solar farm" even though private developers and contractors do much of the early work.

Independent power producer ownership

A huge share of utility-scale solar around the world is owned by *independent power producers*, or IPPs. These companies' business model is to develop, own, and operate power plants and sell electricity under long-term contracts. They don't run the grid; they just supply clean energy into it.

In the IPP model, the developer takes on the work of finding land, securing permits, obtaining interconnection agreements, and arranging financing. After the project is built, the IPP earns revenue by selling power, typically under a Power Purchase Agreement (PPA) with a utility, a large corporate buyer, or a community choice aggregator. These long-term contracts give lenders and investors enough certainty to fund large projects.

Because IPPs compete to offer the lowest-cost energy, this model has driven a lot of innovation and cost reduction in utility-scale solar. Developers look for the most efficient designs, the best equipment pricing, and the strongest sites so they can win competitive procurements. Many of the large-brand-name solar developers you hear about are IPPs or closely tied to them.

Ownership through institutional capital and infrastructure funds

The third major ownership model involves institutional investors and infrastructure funds. These money sources include pension funds, insurance companies, sovereign wealth funds, and specialized infrastructure investment vehicles that want long-lived, predictable cash-flow assets. To them, a solar farm looks a lot like a toll road, a data center, or a water utility — essential infrastructure with relatively steady revenue.

This type of investor often acquires projects after they are operating or nearly complete, rather than developing them from scratch. A developer or IPP might build and stabilize a project and then sell it to an infrastructure fund that plans to hold it for 20 or 30 years. In other cases, funds may co-invest alongside developers from an earlier stage, providing capital in exchange for long-term ownership stakes.

From the grid's perspective, nothing changes when ownership transfers to an infrastructure fund. The project still delivers solar energy under the same power contract. What changes is the entity that receives the cash flows and is responsible for long-term asset management, maintenance contracts, and refinancings. This ownership model has become increasingly common as solar power projects have matured into a mainstream asset class.

Knowing Where You Find Utility-Scale Projects

Utility-scale solar shows up in places where all the puzzle pieces align: strong sunlight, open land, supportive communities, and, above all, access to the power grid. Unlike rooftop systems that can go almost anywhere, large solar farms depend heavily on what happens underground (soil) and overhead (transmission lines). If either of those elements isn't right, the project may never leave the drawing board.

Developers start by evaluating solar resource. Regions with long, bright days and clear skies produce more electricity per acre, which makes the economics stronger. That's why you see a high concentration of utility-scale solar projects in the U.S. Southwest. But growth is accelerating across the Midwest, Southeast, and Northeast as technology improves and costs fall.

Still, sunshine alone isn't enough. Solar farms need a place to plug in, and transmission capacity is often the biggest constraint. It's like having a water source but no pipeline, which means you can't deliver the water to anyone. Proximity to existing high-voltage lines, or the ability to upgrade them affordably, is a make-or-break factor when looking for a location to build a utility-scale solar project. Land near substations or along transmission corridors is especially attractive.

Finally, the human side matters. Successful projects require local support and a plan to avoid or minimize conflicts with environmental habitats, historic landscapes, or agricultural priorities. Some of the most celebrated utility-scale projects sit on former industrial land, alongside ranchland where sheep graze under the panels, or in pollinator-friendly meadows that restore soil health.

Table 13-2 offers a snapshot of these location factors, on which the next sections expand.

TABLE 13-2 **Key Factors for a Utility-Scale Solar Project Location**

Key Factor	Why It Matters	What *Good* Looks Like
Strong solar resource	Better energy output = better economics.	High annual solar irradiance
Transmission access	Delivery of power to the grid is essential.	Proximity to substations or transmission lines with load capacity
Land availability	Quality acreage for panels, spacing, and access roads is necessary.	5–8 acres per MW of power production, minimal slope
Soil and terrain conditions	Conditions impact construction process and cost.	Stable soils, gentle grade, minimal required grading
Community compatibility	Local support means smoother permissions and approvals.	Land uses that align with solar (such as, grazing, low-impact agriculture)
Environmental constraints	Protecting habitats and avoiding long delays is paramount.	Minimal sensitive wildlife species and wetlands impacts

Noting the Attributes of a Good Project

After a developer identifies a region that might support a large-scale solar installation, the real evaluation begins. As I note in the section "Knowing Where You Find Utility-Scale Projects" earlier in the chapter, not every sunny, open parcel of land is a strong candidate for a large solar power system.

REMEMBER

A good utility-scale solar project is one that can produce a lot of power, plug into the grid efficiently, and clear environmental-impact hurdles and community reviews without long delays or costly redesigns. Developers weigh dozens of variables, but the three that can make or break a project most often are solar resource, transmission access, and local compatibility.

High available solar resource

Electricity production depends on local sunlight conditions, the brighter and more consistent the sunshine, the more power a project can deliver each year. Developers rely on long-term NASA and National Renewable Energy Laboratory (NREL) solar resource data to estimate annual energy output. Higher energy production means lower cost per kilowatt-hour, which makes a site more competitive when bidding to sell power.

Figure 13-1 shows a map of solar resource potential globally. This map is available online at the Global Solar Atlas website. To access the map shown, go to the website at `https://globalsolaratlas.info`, click PV Study in the top menu bar, and scroll down.

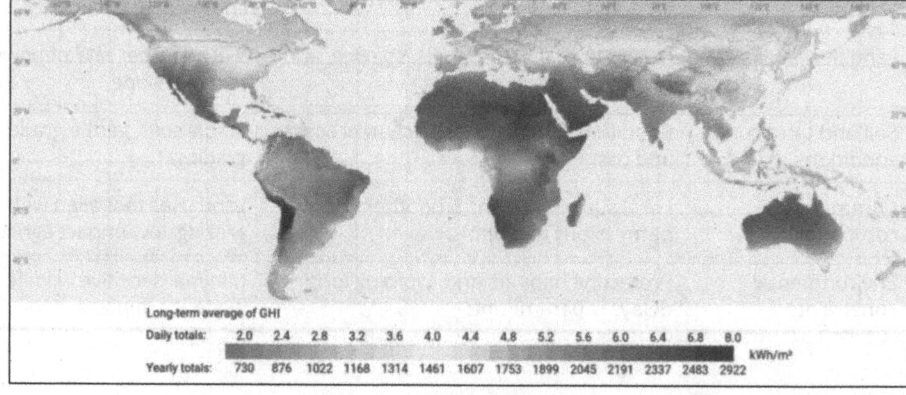

FIGURE 13-1: Map showing global solar resource potential.

Solar resource potential tends to be higher near the equator but also depends on general atmospheric conditions. Even so, utility-scale solar isn't limited to deserts or the U.S. Southwest anymore. Because technology has become more efficient and costs continue to drop, strong projects now make sense in places like the U.S. Midwest, Mid-Atlantic, and New England, even though these locations have somewhat less sunshine.

For a large-scale solar power project location, annual output of the solar resource matters more than the highest single sunny day. Developers prefer a consistent solar resource over extreme peaks.

Proximity to transmission infrastructure

Access to the power grid is often the single most important factor in large-scale solar project success. Utility-scale solar must connect to high-voltage transmission lines — not neighborhood distribution poles — so power can flow across regions. If a site is too far from a substation or if the transmission system is already congested, the project may face expensive upgrades or long delays.

Developers often look for

>> **Existing high-voltage lines** that have available capacity

>> **Nearby substations** that are capable of accepting new power generation

>> *Interconnection queues* (queues for proposed power generation and storage projects waiting for approval to connect to the grid) that aren't too backed up

Interconnection studies (covered in more detail in Chapter 14) determine whether a project can join the grid affordably. Many promising sites are abandoned because the electrical upgrades required for interconnection with the grid would be more expensive than the solar farm itself. Table 13-3 gives you a quick look at the impact of power transmission potential on a large solar project.

TABLE 13-3   **Transmission Impact on Project Viability**

Transmission Potential	Project Outcome
Nearby line with spare capacity	Very attractive site for a viable project
Nearby line that needs upgrades	Possibly viable project if upgrade costs are low
Location remote from transmission lines	Rarely viable project
Congested region with long interconnection queue	Often not a viable project due to timeframes for approvals

Local compatibility and acceptance

Even the best engineering plan won't succeed without local acceptance. Developers must ensure that their project fits with the land and the people who live near it. That means avoiding sensitive habitats, cultural or historic sites, high-value farmland, or areas where the project would dramatically change community character.

Common strategies for proposing a large solar installation site include:

>> Selecting already-disturbed or low-productivity land

>> Using wildlife-friendly fencing and vegetation

>> Partnering with agricultural landowners to allow grazing or farming between rows

>> Early community engagement and transparency

Fewer conflicts between the project and local residents or environmental factors mean faster permitting, stronger community relationships, and reduced long-term legal risk.

A good utility-scale solar project isn't just sunny and connected, it's a project that neighbors want to live with for decades.

Chapter 14

Following the Journey of a Project

Utility-scale solar projects don't appear overnight. What looks like a neat grid of interconnected panels on power-up day is the result of years of planning, negotiation, study, and risk-taking. Before a single panel is installed, developers must secure land, navigate local politics, satisfy environmental rules, convince utilities to accept new power, and line up millions (or billions) of dollars in capital. Many utility-scale projects never make it that far.

This chapter follows a utility-scale solar project from idea to operation. You can see how projects take shape long before construction begins, why interconnection with the grid can make or break a proposed installation site, and how financial and technical decisions ripple through the entire lifecycle. The chapter also explains why the development segment of the project is often called the *hardest and most important* part of the process.

Identifying the Main Player: The Developer

Every utility-scale solar project has one central figure pulling the strings: the developer. Developers are the quarterbacks of the project, coordinating landowners, engineers, utilities, regulators, financiers, and community stakeholders. They take on early-stage risk, invest years of effort before any revenue exists, and decide whether a project is worth pushing forward or walking away from.

A developer's job starts with identifying promising sites, places with good solar resource, feasible grid access, and manageable local constraints (see the related information in Chapter 13). From there, they work to secure land rights, commission technical studies, and begin conversations with utilities and regulators. At this stage, nothing is guaranteed. Developers often spend significant money on projects that are never built because one key piece doesn't fall into place.

What makes developers unique is their willingness to operate in uncertainty. They assemble the project step by step, knowing that interconnection costs may spike, permits can be denied, or financing terms can shift. If the project succeeds, they're rewarded through ownership, development fees, or the eventual sale of the solar power system asset to a long-term owner. If the project fails, they absorb the loss.

Most developers don't build or operate projects themselves. Instead, they focus on creating something financeable: a project with secured land, permits in progress or approved, a viable interconnection path, and a clear route for selling the power produced. After those pieces are in place, developers may bring in partners, sell the project, or transition into a long-term ownership role.

As of this writing, the following list includes companies that are widely recognized as among the largest utility-scale solar developers in the United States, based on operating capacity, development pipeline, or projects brought to market.

AES Clean Energy	NextEra Energy Resources
Cypress Creek Renewables	Orsted Onshore Americas
EDF Renewables America	Recurrent Energy
Invenergy	Silicon Ranch
Intersect Power	

Living Through the Development Lifecycle

Utility-scale solar projects don't spring fully formed from the ground. They are built slowly, deliberately, and often invisibly, through a development process that can take five to ten years before a single panel is installed. This initial phase of the project — the development lifecycle — is where most projects succeed or fail. And success or failure doesn't depend on the capability of the technology but, instead, on the land, people, permits, and power grid requirements that are far more complicated than solar panels themselves.

At a high level, development is about turning a promising idea "this could be a good solar site" into a solid proposal that banks will finance, utilities will interconnect with, and communities will accept. Each phase of the development process builds on or overlaps with the last, and missteps early on can ripple through the project years later.

Before getting into the details, it helps to zoom out and see the full development journey and flow of activities at a glance. Utility-scale solar development is not a straight line, but most projects move through the same broad phases, often with overlap. Some phases happen in parallel, and others stall while a single bottleneck (the interconnection phase, many times) is resolved. Figure 14-1 gives you a high-level look at the development process and how the phases build on each other and overlap throughout the lifecycle.

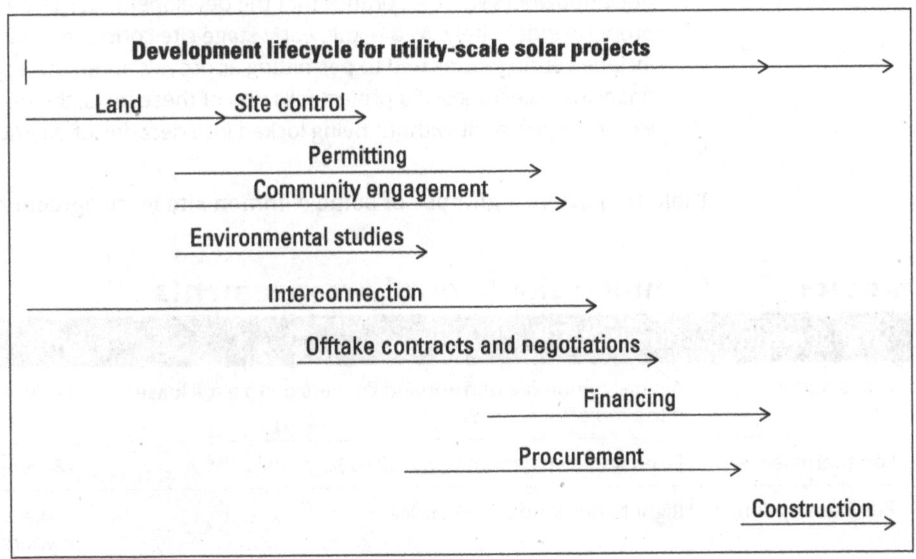

FIGURE 14-1: Development process and key phases.

Land and site control

Every utility-scale solar project begins with a deceptively simple question: Who controls the land? Until that question is answered clearly and legally, everything else about the project can exist only theoretically. Developers can run models, draw layouts, and talk to utilities, but without site control, no project can happen.

Defining site lease timeframes

In utility-scale solar, site control almost always means long-term land agreements, not outright ownership. Developers typically secure leases or purchase options that give them the exclusive right to develop the land for solar power usage over several decades.

Here are some characteristics of site lease agreements:

>> **Multiple timeline terms.** A common site lease structure involves an initial development term (often 3 to 7 years), followed by an operating term of 20 to 35 years after the project is built. The structure may also include optional lease extensions. These timelines are long because solar projects are designed to operate for decades, and lenders require certainty that the land will remain available for the life of the financed asset.

>> **Careful structure that balances risk.** Developers don't want to fully commit capital before they know that a project can be permitted and interconnected, but landowners want assurance that the developer is serious and not tying up property indefinitely. As a result, early-stage site control agreements usually include contingencies tied to permitting approvals, interconnection results, or financing milestones. If a project fails one of these tests, the developer can exit the agreement without being locked into decades of payments.

Table 14-1 offers a glimpse of some common site lease agreement components.

TABLE 14-1   **Common Site Control Arrangements**

Arrangement	How It Works	When It's Used
Lease option	A small option fee upfront and conversion to a full lease if the project advances	Early-stage development
Long-term lease	Fixed annual payments over 20 to 35 years	Most operating projects
Purchase option	Right to buy land at a set price	Projects requiring ownership
Easements	Limited rights for site access or transmission	Supporting infrastructure

Vetting and reclaiming the land

The type of land matters just as much as the lease agreement itself. Utility-scale solar often sits on agricultural land, marginal farmland, former industrial sites, or other large, open parcels. Developers assess not only acreage, but also topography, soil conditions, access roads, drainage, and proximity to neighbors. A site that looks perfect on a map may turn out to be evaluated as too steep, flood-prone, or fragmented when engineers walk the land in person.

Landowners also vary widely in their motivations. Some see solar as a stable, long-term income stream that complements farming or ranching. Others are primarily concerned about preserving future flexibility or ensuring the land is restored at the end of the project's life. To address these concerns, most leases include decommissioning provisions that require the developer to remove equipment and restore the land when the project is retired.

A solar project doesn't typically "own" the land; it borrows it for a generation and must give it back responsibly.

Securing a site at the right time

From the developer's perspective, land control is also about timing. Securing a site too late risks losing it to a competing developer. Securing it too early means carrying legal costs and relationship management for years before revenue from a solar power installation appears. Experienced developers spend significant effort to sequence land control arrangements with permitting, environmental studies, and interconnection so that no single phase of the development project gets too far ahead of the others.

Land and site control phases of a solar development process rarely make headlines, but they quietly shape everything that follows. A well-structured site agreement gives developers flexibility, reassures investors, and sets expectations with landowners from day one. A poorly structured one can derail a project years later, just when it looks ready to move forward.

Permitting and local approvals

After land is secured, the project steps into the public eye and into the world of local rules, hearings, and approvals. If land control makes a solar project legally possible, permitting is what makes it publicly acceptable. At this stage, a project moves from a private agreement between a developer and a landowner into something the surrounding community can see, question, and influence. For many projects, the permitting phase can cause timelines to stretch and uncertainty to grow.

Utility-scale solar projects are subject to a patchwork of local, state, and sometimes federal rules. These rules govern how developers can use the land, what structures they can build, how they must manage water resources, and how the projects can interact with neighboring properties. In some jurisdictions, a solar power installation becomes *by-right use* on certain land types, meaning the project can proceed with minimal approvals. In other jurisdictions, a solar installation project can trigger discretionary reviews that require public hearings, planning commission votes, or even county board approval.

Focusing on approvals, hearings, and permits

Local approvals for solar power projects often focus on practical concerns such as setbacks from property lines, height limits for racking systems, glare impacts, fencing and access roads, stormwater management, and emergency vehicle access. None of these requirements are especially controversial on their own, but together they shape the project's layout and cost. A slightly wider setback or a less suitable access road location can remove acres from the usable site; this loss of usable land, in turn, reduces energy output and alters project economics.

Public hearings are a defining feature of the permitting and approvals phase. They provide a formal forum for neighbors and stakeholders to ask questions, raise objections, or support the project. Concerns commonly include visual impact, effects on property values, congestion from traffic during construction, and long-term land use. Skilled developers treat these meetings as opportunities to explain the project clearly and adjust designs where possible, rather than as obstacles to be rushed through.

Permitting timelines are rarely predictable. A straightforward project in a solar-friendly jurisdiction may receive approvals in under a year. A similar project elsewhere might take several years, especially if the project location requires zoning changes or conditional-use permits. Appeals and legal challenges, while not common, can add further delays.

Coordinating permits across agencies and organizations

Developers must also coordinate permits across agencies and organizations. Local approvals often interact with environmental requirements, utility interconnection timelines, and financing milestones. Table 14-2 gives you a look at common local permits for a utility-scale solar project. A delay in obtaining one permit can cascade the approval phase of the project into other milestones and force developers to renegotiate contracts or update studies that have gone stale.

  Common Local Permits for Utility-Scale Solar Projects

Permit Type	What It Covers	Who Issues It
Zoning or land-use approval	Whether a solar installation is allowed on the site	City or county
Site plan approval	Layout, access roads, drainage	Planning commission
Building permits	Structural and electrical safety	Local building department
Stormwater permits	Runoff and erosion control	Local or state agency
Conditional use permit	Case-by-case approval	Local governing body

The permitting phase of development can feel slow because it is designed to be deliberate. It forces projects to balance regional energy goals with local priorities, and although that tension can be frustrating, it also shapes projects that are more durable over time.

Environmental studies

After permits are underway, the project turns its attention to understanding the land itself — not just legally, but environmentally. Gathering environmental information helps developers understand the land and also informs the permitting phase. These two phases of project development shape each other over time. Environmental studies can feel like a bottleneck, but they serve an important purpose. They help ensure that solar projects deliver clean energy without creating new problems, and they often strengthen a project's long-term resilience.

Even though solar power is widely viewed as a low-impact form of energy, utility-scale projects still need to prove that they won't cause unintended environmental harm. Conducting and documenting environmental studies are how developers demonstrate that a project fits responsibly into its surroundings.

What environmental studies study

Environmental studies look closely at the characteristics that already exist on or near a proposed solar site: wetlands, streams, wildlife habitat, agricultural soils, historic or cultural resources, and nearby protected areas. The goal of assessing the site location isn't perfection (no construction project can have zero impact); it's understanding what effects the project will have. Regulators want to know how a project might change drainage patterns, disturb sensitive species, or alter land that has other recognized value.

Environmental review requirements vary widely depending on location and jurisdiction. Some projects require limited studies focused on wetlands and stormwater, while others trigger comprehensive state or federal reviews. In certain cases, projects that involve federal land, federal funding, or federal permits must comply with national environmental review laws, adding additional layers of documentation and public review.

Timing matters enormously when conducting studies. Many environmental surveys can happen only during specific seasons. Wildlife studies may require spring or summer fieldwork, while plant surveys may depend on particular bloom periods. If your study misses a season, your project may be delayed by months, sometimes an entire year, waiting for the next survey window.

Informing system design and building trust

Developers don't conduct environmental studies simply to satisfy regulators. The study results can directly inform project design. For example, a system designer may adjust array layouts to avoid wetlands, expand buffer zones near sensitive habitats, and reroute access roads to protect waterways. These changes can reduce the total buildable area, but they often make the difference between a project that moves forward smoothly and one that becomes mired in opposition or legal challenges.

Environmental reviews for a solar power project can also play an important role in building trust. Transparent studies give communities confidence that system developers have considered their project's impacts carefully. When developers can point to independent assessments and mitigation plans, discussions with local officials and neighbors tend to be more constructive.

Table 14-3 shows some common environment studies performed for utility-scale solar projects.

TABLE 14-3   **Common Environmental Studies for Utility-Scale Solar Projects**

Study Type	What It Evaluates	Why It Matters
Wetlands delineation	Presence of regulated wetlands	Determines setbacks and land exclusions
Wildlife surveys	Birds, bats, or protected species	Avoids harm to wildlife and legal risk from regulatory violations
Cultural resources	Historic or archaeological sites	Protects heritage and avoids delays
Stormwater analysis	Runoff and erosion	Prevents flooding and water pollution
Agricultural impact	Soil quality and land use	Guides restoration and coexistence plans

Interconnection: The true hero

After developers understand any environmental constraints, the project faces its most formidable challenge: connecting to the grid. If there is one phase of utility-scale solar development that deserves more respect than it gets, it is interconnection. Panels, land, permits, and financing don't matter if a project can't physically deliver power to the grid. Interconnection determines not only whether a project can move forward, but when, where, and at what cost.

Interconnection is the process by which a solar project applies to connect to the existing electrical grid and undergoes a series of technical studies to assess impacts. These studies evaluate whether nearby transmission or distribution infrastructure can safely handle the additional power, and what upgrades, if any, are required to make interconnection possible. In theory, this phase sounds straightforward. In practice, it is often the most uncertain, expensive, and time-consuming part of solar power system development.

Interconnection is often called the "true hero" of a utility-scale solar project — not because it is glamorous — but because it quietly determines success. Projects that navigate interconnection well move forward smoothly. Projects that underestimate it often stall indefinitely.

What makes interconnection especially challenging is that it is not a static process. Here's what happens:

>> **Projects typically enter an interconnection queue managed by a utility or regional grid operator.** After projects are in the queue, they wait their turn for studies that assess thermal limits, voltage impacts, short-circuit capacity, and system stability. Each phase of study builds on the last, and unfavorable results at any stage can force costly redesigns or even project abandonment.

>> **The grid changes as new power generators enter the queue, existing plants retire, and demand patterns shift.** A project that looked viable when it applied to enter the interconnection queue may face higher upgrade costs years later simply because other projects moved ahead of it in the queue.

Upgrade costs are the defining risk for this phase of solar project development. In some cases, interconnection requires only minor equipment additions. In others, it may trigger multi-million-dollar transmission upgrades that make the project economically unfeasible. Developers must decide whether to absorb the upgrade costs, redesign the project, relocate it, or walk away entirely.

Table 14-4 offers a quick look at typical outcomes of the interconnection phase of solar project development.

Typical Interconnection Outcomes

Interconnection Result	Impact on Project
Minimal upgrades needed	Project advances quickly
Moderate upgrades needed	Economics may need to be reworked
Major transmission upgrades	Often fatal to the project
Queue delays or restudies	Timeline stretches years

Community engagement

After developers understand the interconnection situation — even if the outcome isn't ideal and cost-free — they can engage more confidently with the affected communities and know what kind of project is realistic for that location. Even when a project has secured land, with permits underway, environmental studies complete, and interconnection looking feasible, it still faces one final test: whether the surrounding community is willing to live with it. Community engagement isn't a box to check at the end of development; it's a thread that runs through the entire process and often determines how smoothly everything else unfolds.

Utility-scale solar projects can be unfamiliar, especially in rural or semi-rural areas where large infrastructure projects are rare. Neighbors may have questions about what the site will look like, how long construction will last, whether property values will change, or what happens to the land decades from now. Some concerns are practical; others are emotional; and both matter.

Engaging with all local communities early

Developers who engage with local communities early in the development process tend to fare better than those who wait until formal hearings force the conversation. Early outreach allows developers to explain the solar installation project in plain language, listen to feedback, and make adjustments before opposing positions harden. Changes may include elements such as increasing setbacks, improving vegetative screening, adjusting access routes, or clarifying decommissioning plans. And making these changes can significantly reduce opposition without harming project economics.

Community engagement also extends beyond neighbors. Local officials, planning staff, emergency responders, and agricultural stakeholders all need to understand how the solar power project will operate. Clear communication helps align expectations and prevents misinformation from filling the gaps.

Communities are more receptive when they feel consulted about project plans, not informed after decisions are already made.

Expressing long-term benefits

Many developers focus on presenting long-term benefits and not just asking for approval. Utility-scale solar power projects often provide stable tax revenue for local governments, lease income for landowners, and temporary construction jobs. Some projects include community benefit funds, educational partnerships, or commitments to local hiring. While these don't eliminate opposition entirely, they can shift the conversation from "What is this project taking?" to "What is this project providing?"

Community engagement doesn't end after construction begins. Solar plants operate quietly for decades, and developers or owners remain visible members of the community. Projects that start with transparency and follow through on commitments tend to avoid conflicts later, especially when equipment needs maintenance or upgrades.

EPC selection and procurement

As a project nears the end of its development lifecycle, attention shifts from *whether* it can be built to *how* it will be built. In this phase, EPC selection and material procurement come into play. EPC stands for engineering, procurement, and construction, and the EPC contractor is responsible for turning years of planning into a functioning solar power plant.

Here are the two main decisions that developers make during this development phase:

>> **Selecting an EPC partner.** The choice of a partner is one of the most consequential decisions a developer makes. EPCs don't just build projects; they lock in costs, schedules, equipment choices, and performance guarantees. A strong EPC contract can reduce risk and increase confidence among lenders and investors. A weak one can introduce uncertainty that ripples through financing and long-term operations.

>> **Procuring the major equipment.** Sourcing equipment for a solar power system includes securing components such as *modules* (the solar panels), inverters, transformers, and racking to hold the other components. Selecting equipment sources and methods is tightly intertwined with EPC selection. Developers must decide whether to let the EPC handle procurement entirely or to lock in certain components themselves. These decisions affect pricing, warranties, delivery timelines, and exposure to supply-chain disruptions.

Timing is critical during this project phase. Developers often begin EPC conversations well before financing closes by using preliminary pricing and technical assumptions to refine project economics. In some cases, they must reserve equipment or order it early to meet interconnection or commercial operation deadlines. This timing can mean that developers are committing capital before financing is fully secured, which is another example of how development risk concentrates near the end of the lifecycle.

EPC contracts typically define responsibilities for engineering design, construction management, quality control, testing, and commissioning. They also specify performance guarantees, liquidated damages for delays, and remedies if the system underperforms. These provisions are closely scrutinized later by lenders and investors, which is why EPC selection is inseparable from financing readiness.

EPC decisions don't just affect construction, they shape financing, schedules, and long-term project performance.

REMEMBER

Addressing Financing and Capital Stack

By the time a solar power system developer selects an EPC and defines procurement strategies, a project has crossed an important threshold. It has moved from planning into execution. At this point, the remaining question is no longer "Can this project be built?" but "Will the numbers support the project?"

At utility scale, solar power system development eventually stops being about land, permits, studies, and relationships and starts being about money. Financing a project defines the moment when a project either becomes real or quietly falls apart. Lenders and investors don't care how elegant a site map looks or how passionate a developer is; they care whether the project can reliably generate cash for decades.

The combination of funding sources that make a project work is known as the *capital stack*. Each layer in the stack plays a different role, takes a different kind of risk, and expects a different return. Understanding how these layers fit together helps explain why developers make certain design decisions, why timelines matter so much, and why projects sometimes change shape late in the development cycle.

Most utility-scale solar projects rely on some mix of equity, debt, and tax equity tied together by detailed financial models that test whether the project is *bankable* (in this case, whether it's a reliable investment).

Equity

Equity is the foundation of the capital stack. Equity investors provide upfront capital in exchange for ownership in the project and a share of its long-term returns. This money is at the greatest risk because if the project underperforms or fails, equity investors are the first to absorb losses — but being in the equity piece of the capital stack also has the greatest upside.

In early development, equity funds often come from the developer or from strategic partners willing to take development risks. As projects mature and move closer to construction, long-term owners such as infrastructure funds, pension funds, or renewable energy platforms may step in to provide additional equity. Chapter 13 has more information about project ownership.

Equity investors care deeply about risk management. They scrutinize interconnection results, permitting status, equipment warranties, and operating assumptions. These investors may still offer financing to a solar power project that looks slightly riskier, but only if returns appear high enough to compensate for the risk involved.

Debt

Debt financing usually arrives after a project is far enough along that its risks are understood and controlled. Debt financing comes from banks or other lenders and is typically repaid over time by using the project's operating revenue.

Because lenders in the debt layer are paid back before the equity layer receives profits, debt is cheaper than equity but also more conservative. Lenders require strong contracts, predictable cash flows, and technical designs that minimize surprises. They care less about upside and far more about downside protection.

TECHNICAL STUFF

Debt terms are heavily influenced by project fundamentals: expected energy output, power purchase agreement terms, operating costs, and reserve requirements. Even small changes to assumptions can affect how much debt a project can support.

Tax equity

Tax equity is a uniquely important layer in U.S. utility-scale solar financing. Because utility-scale solar projects generate federal tax benefits — including investment tax credits and accelerated depreciation — developers often partner with investors who can use those benefits efficiently.

Tax equity investors are typically large financial institutions with significant tax liability. They invest capital upfront in exchange for most of the project's tax benefits and a portion of early cash flows. Over time, ownership and cash distributions revert back toward the developer or long-term owner.

Tax equity structures are complex and highly sensitive to policy, accounting rules, and project timing. Delays in construction or changes to equipment eligibility can materially affect tax equity value, which is why developers coordinate closely with tax advisors throughout development.

Financial modeling and bankability

All of the layers in the capital stack for a solar power project come together in financial models. These models project energy production, revenue, expenses, taxes, debt service, and investor returns over the life of the project. The project developers and lenders update the model constantly as real data replaces assumptions on performance forecasting, equipment pricing, financing terms, and market conditions.

A project is considered bankable when lenders and investors agree that its risks are understood, mitigated, and appropriately priced. *Bankability* doesn't mean risk-free; it means predictable, with technologies that are proven, contracts that are enforceable, and counterparties that are credible.

Decisions made earlier in development — such as site selection, equipment choice, and interconnection strategy — echo loudly here. A slightly cheaper inverter or a marginal site layout might look attractive early on, but if it introduces uncertainty, it can raise financing costs or kill a deal entirely.

Bankability isn't about maximizing performance, it's about minimizing surprises.

Operations and Maintenance at Utility Scale

After a utility-scale solar project is built and commissioned, its success depends on how well the solar power plant operates and is maintained over time. Unlike residential solar systems, which often run quietly with minimal oversight, utility-scale projects are actively managed assets. They are expected to perform predictably every day for decades, and small performance issues can add up to meaningful financial losses at scale.

Operations and maintenance, usually shortened to O&M, covers everything required to keep a solar plant producing energy safely and reliably. This task includes monitoring system performance, inspecting equipment, responding to outages, maintaining vegetation, cleaning modules when necessary, and coordinating repairs. O&M providers may be part of the project owner's organization or a specialized third party contracted to manage the plant. Table 14-5 contains information about typical O&M responsibilities.

TABLE 14-5

Typical O&M Responsibilities at Utility Scale

Activity	Purpose
Performance monitoring	Detects underperformance quickly
Preventive inspections	Reduces long-term failures
Vegetation management	Prevents shading and access issues
Corrective maintenance	Restores output after faults
Safety compliance	Protects workers and equipment

Here are two important features of O&M oversight:

>> **At utility scale, power production monitoring is continuous.** Modern plants use sophisticated software platforms that track energy output, equipment status, and grid conditions in real time. If an inverter goes offline or production drops unexpectedly, the software triggers an alert immediately. Fast response matters because a single inverter outage at a large plant can mean megawatts of lost generation.

>> **Preventive maintenance is just as important as reactive fixes.** Regular inspections of the plant's equipment help catch issues — such as loose connections, degraded components, or shading from vegetation growth — before they become system failures. Because projects are designed to operate for 25 to 40 years, consistent maintenance protects both performance and asset life.

Good O&M keeps solar power production close to forecasts, which is critical for meeting contractual obligations under power purchase agreements and for maintaining lender and investor confidence.

Providing for Long-Term Asset Management and Revenue Optimization

Operating a solar plant is about more than keeping the lights on. Over a multi-decade lifespan, owners focus on asset management and revenue optimization, ensuring that the project delivers the financial returns it was designed to provide.

Asset management sits above day-to-day O&M (see the preceding section) for a utility-scale solar power plant. It includes managing contracts, tracking warranties, overseeing compliance with financing agreements, coordinating insurance, and planning long-term system upgrades or replacements. Asset managers act as the project's stewards, making sure that nothing quietly erodes value over time.

Revenue optimization involves maximizing the solar project's earnings within its contractual and grid constraints. Tasks related to revenue optimization include verifying energy payments, managing *curtailment events* (periods when grid operators require the project to reduce output due to grid congestion or oversupply), optimizing inverter settings, and increasingly, coordinating solar output with storage or market signals. Even small percentage improvements in power availability or delivery efficiency can translate into significant revenue gains over the life of a large project.

Long-term owners must also plan for change. Inverters may be replaced, monitoring systems upgraded, and repowering decisions evaluated as technology improves. Near the end of a project's life, asset managers assess whether extending operations makes sense or whether decommissioning and site restoration should proceed.

The most profitable utility-scale solar power projects are rarely the flashiest; they're the ones managed consistently and conservatively over their lifetimes.

Chapter **15**

Finding Out Who Buys the Power

Utility-scale solar power installations don't exist in a vacuum. Unlike rooftop systems, which serve a single home or business, large solar projects are built to sell power into complex markets governed by contracts, regulations, and grid rules. Panels may generate electricity reliably, but whether a project succeeds financially depends on who agrees to buy that power, under what terms, and for how long.

This chapter focuses on the commercial side of utility-scale solar: the buyers, the contracts, and the market structures that turn electrons into revenue. You find out why long-term agreements are so important, how various buyers approach solar procurement, and what happens when projects operate without guaranteed buyers at all. You also discover how policy choices and grid operations shape the risks developers and owners take on, often in ways that aren't obvious from the outside.

In any case, this chapter gives you an understanding of why the question "Who buys the power?" is often the first one that financiers ask, and how the answer determines everything from project size to location to long-term value.

Understanding Power Purchase Agreements

At utility scale, solar power plants are rarely built on speculation alone. Before lenders commit capital or construction begins, some entity usually needs to agree to buy the electricity the project will produce. That contractual agreement is imaginatively called a *power purchase agreement* (PPA), and it sits at the center of most utility-scale solar economics.

Seeing the scope of power purchase agreements

A PPA is a long-term contract between a power producer and a buyer that spells out how much electricity will be sold, for how long, and at what price. These contracts often have a term of 15 to 30 years, which provides revenue certainty that makes large infrastructure investments (like solar power plants) possible. Without a PPA, a solar project may still exist, but it becomes much riskier, both financially and operationally.

The PPA defines pricing structures, delivery obligations, performance standards, and remedies if the project underperforms or fails to deliver energy as expected. For developers and financiers, these details matter just as much as sunshine and equipment. In its simplest form, a PPA answers three questions:

>> **Who buys the power?** Because PPAs are long-term commitments, buyers are careful about who they contract with. Creditworthiness matters. A buyer with strong financial backing provides confidence that payments will arrive reliably over decades. This is why utilities, large corporations, and government-backed entities are common counterparties in utility-scale PPAs.

>> **How much will the buyers pay?** Most PPAs are structured around a fixed price per megawatt-hour of electricity delivered to the grid. That price may escalate slightly over time or remain flat for the life of the contract. The predictability of these payments is what allows banks to model cash flows decades into the future and determine how much debt a project can support.

>> **And what happens if something goes wrong?** PPAs also allocate risk. They define who bears the cost of grid outages, *force majeure events* (extraordinary, uncontrollable occurrences such as war or natural disaster), *curtailment* (intentional reduction of output), or regulatory changes. For example, if the grid operator orders a plant to reduce output, the PPA determines whether and how much the buyers still pay into the project. These provisions can significantly affect project value, even if they rarely make headlines.

In the realm of utility-scale solar power, a PPA doesn't just arrange to sell electricity, it enables the infrastructure project to exist.

Here are a few key elements of a utility-scale PPA:

>> **Contract term:** Defines the length of time that the agreement covers

>> **Price structure:** Establishes a fixed, escalating, or indexed and specific amount, usually in dollars per megawatt hours ($/MWh)

>> **Delivery point:** Details the connection where power is injected into the grid

>> **Performance obligations:** Describes the expected timeframe for availability of power and the output amount

>> **Curtailment provisions:** Explains the management of payments and outlines conditions related to grid restrictions

>> **Force majeure:** Describes the handling of the agreement in the case of uncontrollable events

Recognizing the unique elements

PPAs are negotiated documents, not standardized forms. While many provisions are common across the industry, specific contract terms vary based on market conditions, project location, buyer preferences, and regulatory context. In competitive project procurement processes, developers often bid aggressively on price to win the buyers request-for-proposals (RFP) process to supply electricity but seek flexibility in other terms — such as delivery schedules, curtailment risk allocation, and performance guarantees — to manage their risk.

You should also understand what PPAs don't do. A PPA does not guarantee profits. It guarantees a buyer, not favorable economics. If construction costs rise, equipment underperforms, or operating expenses exceed expectations, the PPA price doesn't automatically adjust. That's why developers and investors scrutinize the terms of these contracts so closely.

Over time, PPA structures have evolved. Early utility-scale solar relied almost exclusively on utility buyers. Today, a broader mix of buyers, including corporations and community energy providers (such as Community Choice Aggregators, or CCAs), participate in long-term contracts. Some projects even operate without PPAs at all, accepting market risk in exchange for potential upside.

When agreements go virtual

You should be aware that not all PPAs involve the physical delivery of electricity to specific buyers. A virtual power purchase agreement (virtual PPA, or VPPA) is a financial contract that involves the broader power market rather than a physical delivery to the buyers. In a VPPA, the solar project sells its electricity into the wholesale market, while the buyer and the project settle the difference between the market price and a fixed contract price. If market prices are higher than the VPPA price, the project pays the buyer; if they're lower, the buyer pays the project.

In a VPPA, the buyer doesn't receive electrons directly, but they do receive the environmental attributes of providing cleaner energy and price certainty. These attributes make VPPAs especially popular with large corporations seeking long-term clean energy commitments without the responsibility of owning power generation or the restriction of operating in a specific location.

Meeting the Power Buyers

While the mechanics of a PPA (see the earlier section "Understanding Power Purchase Agreements") may look similar on paper, who is buying the power makes a big difference in how a utility-scale solar project is structured, financed, and operated. In today's market, most long-term solar contracts fall into three broad buyer categories: traditional utilities, large corporations, and Community Choice Aggregators (CCAs). Each buyer group uses PPAs but comes to the solar power market with different motivations, constraints, and risk tolerances.

Understanding these buyer types helps explain why solar projects can look so different from one another, even when they use similar technology. Check out Table 15-1 for a comparison between buyer types.

TABLE 15-1   **A Comparison of Utility-Scale Solar Power Buyers**

Buyer Type	Who They Are	Why They Buy Solar
Utilities	Investor-owned, municipal, or cooperative electric utilities	To meet demand reliably, comply with regulations, and lock in long-term pricing
Corporations	Large companies (tech, manufacturing, retail, data centers)	To meet sustainability goals, hedge energy costs, and claim clean energy attributes
Community Choice Aggregators (CCAs)	Public entities procuring power for cities or counties	To provide local clean energy and price stability

Across all three buyer types, one theme remains constant: Long-term contracts reduce uncertainty. Whether the buyer is a utility, a Fortune 500 company, or a regional energy aggregator, a signed PPA transforms a solar project from a technical asset into a financial one.

Utilities: The original buyers

Utilities were the original buyers of utility-scale solar power, and they remain the largest purchasers by volume of energy bought. Investor-owned utilities, municipal utilities, and electric cooperatives all use PPAs to procure solar energy as part of their long-term resource planning.

For utilities, buying solar power is primarily about meeting electricity demand reliably and affordably while complying with regulatory requirements. Many utilities operate under mandates to source a certain percentage of their power from renewable energy, and long-term solar PPAs help them meet those goals with predictable pricing.

Utility PPAs tend to be conservative. Contracts often have long timeframes, pricing that is tightly scrutinized by regulators, and clearly defined performance requirements. Utilities value reliability and creditworthiness above all else, which makes them attractive counterparties from a financing perspective. Lenders generally view utility-backed PPAs as among the lowest-risk revenue streams available in the power sector.

That said, procurement that involves utility buyers can be slow. Competitive solicitations like requests for proposals (RFPs), regulatory approvals, and public oversight mean projects may take years to move from bid to contract execution.

Large corporations: The emerging buyers

Over the past decade, large corporations have emerged as major buyers of utility-scale solar power. Technology companies, manufacturers, retailers, and data center operators increasingly use PPAs to meet sustainability commitments, hedge energy costs, and demonstrate leadership on clean energy.

Corporate PPAs often take the form of virtual PPAs (VPPAs, see the earlier chapter section "When agreements go virtual"), which allow companies to support new solar projects without needing physical delivery of power to their facilities. These agreements are attractive because they provide long-term price certainty and environmental attributes while remaining flexible across geographies.

Compared to utilities, corporations may be more willing to accept market exposure or novel contract structures, but they are also highly sensitive to accounting treatment, reputational risk, and contract complexity. Credit quality matters here as well; projects backed by well-known corporate buyers are typically easier to finance than those tied to smaller or less-established companies.

Corporate procurement has expanded the universe of viable solar projects, especially in regions where utilities are slow to contract or where renewable mandates are limited.

Community Choice Aggregators: The local buyers

Community Choice Aggregators, or CCAs, are a newer but rapidly growing category of solar power buyers, particularly in states such as California. CCAs are public or quasi-public entities that procure electricity on behalf of cities or counties while existing utilities continue to handle transmission and billing.

CCAs often prioritize local clean energy, price stability, and community benefits. Many actively seek long-term solar power contracts to support regional projects and meet aggressive clean energy targets. Because CCAs are accountable to local governments rather than shareholders, they may place additional emphasis on environmental performance and local economic impact.

From a project perspective, CCAs can be excellent buyers, but they vary widely in size, financial structure, and experience. Some are well-established with strong balance sheets; others are newer and more sensitive to market volatility. Developers and lenders carefully evaluate each CCA's creditworthiness, structure, and oversight before relying on them as long-term counterparties.

Looking at Merchant Projects and Market Exposure

Not every utility-scale solar project has a long-term buyer locked in. Some projects move forward without a PPA and just sell their electricity directly into wholesale power markets. These situations are known as merchant projects, and they represent a very different risk-and-reward profile from contract-backed solar power production.

Merchant projects are not inherently better or worse than contracted projects; they are simply different. They require that the project owners have stronger balance sheets, greater tolerance for uncertainty, and more sophisticated market strategies. For many investors, especially those seeking steady, long-term returns, PPAs remain the preferred path. For others, merchant exposure offers an opportunity to capture value where contracts are scarce or markets are evolving.

Seeing a merchant project at work

In a merchant project, the solar plant sells power at whatever the market price happens to be at the time the electricity is generated. Prices can change hourly based on supply, demand, weather, fuel costs, and grid conditions. When prices are high, merchant projects can earn more than they would under a fixed-price PPA. When prices are low, revenue drops — sometimes dramatically.

Because the revenue stream is uncertain, merchant projects carry much higher financial risk. Lenders are generally reluctant to finance projects without contracted revenue, which means merchant projects rely more heavily on equity and often use less debt (see Chapter 14 for an overview of solar project financing). That higher cost of capital is the tradeoff developers accept in exchange for potential upside of selling power at higher prices.

Merchant projects don't trade certainty for failure because they can keep producing and selling power. But they do trade certainty for price volatility.

REMEMBER

Planning for market exposure

Merchant project exposure is more common in certain power markets than in others. Regions with deep, liquid wholesale power markets and transparent pricing make merchant strategies more feasible. Developers may also pursue partial merchant exposure, combining contracted revenue for part of a project with market sales for the remainder, or using shorter-term hedges instead of long-term PPAs.

TECHNICAL
STUFF

Some developers intentionally build merchant-ready projects to take advantage of anticipated future conditions. For example, if a region expects rising power demand or transmission constraints, developers may bet that prices will increase over time. This approach requires strong market insight and a willingness to weather lean years.

Accommodating operational risks

Merchant projects also face operational pressures beyond those experienced by other power-buying arrangements. Because revenue depends on market prices, timing matters. Curtailment events (see the section "Seeing the scope of power purchase agreements" earlier in the chapter for more information about these events), grid congestion, or midday oversupply can all reduce earnings. In regions with high solar penetration, prices may drop sharply during sunny hours, a phenomenon sometimes called *price cannibalization*.

To manage these risks, some merchant projects pair solar power generation with energy storage, allowing owners to shift power delivery to higher-priced hours. Others use financial hedges or short-term contracts to smooth revenue. These tools don't eliminate risk, but they can make merchant exposure more manageable.

Recognizing Solar Power Policy Drivers

Even though utility-scale solar projects operate in competitive power markets, governing energy policies still play a quiet but powerful role in shaping where solar power projects are built and who buys the power. These policies don't usually dictate outcomes directly; instead, they set the rules of the game, which influences demand for clean energy, contract structures, and long-term investment decisions.

Policy also affects how power markets operate. Grid rules, interconnection standards, and market design determine how power buyers compensate solar plants and how owners handle curtailment. In some regions, policy reforms have expanded opportunities for solar to participate in capacity markets or provide ancillary services, while in others, outdated rules limit solar power's value.

Policies that govern solar power production don't pick winners, they shape the environment in which the solar projects compete.

Policies that govern solar projects appear at various levels:

>> **At the state level,** renewable portfolio standards (RPS) and clean energy standards require utilities or energy providers to source a certain percentage of their electricity from renewable or zero-carbon sources. These mandates create steady demand for solar power and encourage utilities and CCAs to sign long-term PPAs. In states without such standards, solar projects may rely more heavily on corporate buyers or merchant exposure.

>> **At the federal level,** policies tend to influence solar indirectly by shaping project economics rather than dictating procurement. Incentives such as tax credits and accelerated depreciation improve returns, making projects more competitive in wholesale markets and in utility solicitations. The details of these incentives change over time, but their broader effect is to reduce risk and attract capital.

>> **At the local level,** policies matter, too. Zoning rules, siting guidelines, and permitting processes can either enable or inhibit the solar project development process. Jurisdictions that provide clear, consistent rules tend to attract more projects, but uncertain guidelines can push developers to site projects elsewhere.

Policy drivers don't guarantee success, but they tilt the playing field. Understanding them helps explain why solar grows rapidly in some regions and more slowly in others — even when the amount of sunshine is similar.

Grasping Grid Integration and Curtailment

No matter who buys the power or what contract structure they use, every utility-scale solar project ultimately answers to the grid. Grid integration is the practical reality of delivering electricity into a system that must balance supply and demand in real time. *Curtailment* (intentionally reducing power production) is what happens when the system can't maintain that balance, and it's an increasingly important part of the solar power story.

Managing grid integration

Grid operators manage power flows to keep the entire system stable. When more available electricity exists than the grid can safely use or transmit, operators may instruct generators to reduce output. For solar projects, this means producing less energy than they are physically capable of generating, even when the sun is shining.

Grid integration also involves technical requirements. Solar plants must meet grid codes related to voltage support, frequency response, and ride-through capabilities during disturbances. Modern inverters allow solar plants to behave more like traditional power plants, providing grid services that improve reliability.

Dealing with curtailment

Curtailment is not a sign that a solar power project has failed. In fact, it often reflects success when high levels of renewable generation meeting or exceeding demand during certain hours. But for project owners, curtailment directly affects revenue. Energy that isn't delivered also isn't paid for — unless a contract specifically provides compensation.

Different markets handle curtailment differently. Some PPAs treat curtailment as a shared risk, while others place it squarely on the project owner. Merchant projects bear the full impact. As solar project penetration increases, curtailment becomes a factor that developers must model and manage from the earliest stages of a project. Table 15-2 gives you a look at common curtailment drivers and responses.

TABLE 15-2 **Common Curtailment Drivers and Responses**

Driver	Why It Happens	Typical Mitigation Strategy
Midday oversupply	High solar output with low power demand	Storage to hold excess output, shifting electricity use from when its abundant to when it's in demand
Transmission congestion	Limited grid capacity	Grid upgrades, relocation of connection points
System reliability events	Emergency balancing of supply versus demand	Improved grid services
Market price collapse	Excess power generation	Contract protections

To manage curtailment risk, developers and owners increasingly look to solutions such as energy storage, improved forecasting, and strategic siting near load centers or transmission upgrades. Storage allows projects to shift energy to higher-demand hours, while better forecasting helps grid operators plan more effectively.

5

Seeing What's Next for Solar: The Future Is Bright

Chapter **16**

Establishing Space-Based Solar Power

For decades, solar power has meant panels on rooftops and in fields and deserts, quietly turning sunlight into electricity right where it lands. Space-based solar power turns that association inside out. Instead of waiting for sunlight to reach Earth naturally, space-based systems can place solar infrastructure in orbit and then deliver energy back to Earth in new ways that may potentially extend solar power availability beyond daylight hours and weather limitations.

Space-based solar power asks a simple question: What if humans could manipulate where and when sunlight shows up as a power source? Some concepts rely on orbital mirrors that redirect sunlight to collectors on the ground. Others convert sunlight into electricity in space and transmit that energy wirelessly, either as microwaves or as tightly focused infrared light aimed at solar facilities on Earth. Each approach comes with its own advantages, technical hurdles, and (many) unanswered questions.

This chapter doesn't assume space-based solar is either inevitable or impossible. Instead, it lays out what these systems are, how they differ, and why they keep resurfacing in serious energy discussions. And chapter coverage offers you an understanding of how space-based solar power production may work, where it may add real value, and what challenges developers still need to solve.

Describing Space-Based Solar Power

Space-based solar power refers to a family of concepts that place solar infrastructure in orbit around the Earth and use it to deliver energy to the Earth. What makes these ideas distinct from traditional solar power production isn't the source of energy — it's still the Sun — but the location of the energy-collecting hardware and the added control that location provides. By operating above clouds, weather systems, and the day–night cycle on Earth, space-based solar power systems aim to make solar energy more predictable and more flexible.

All space-based solar concepts share three common elements. They

>> **Rely on satellites positioned in orbit around the Earth,** often in geostationary or sun-synchronous paths, where exposure to sunlight is far more consistent than on the ground.

>> **Use large-scale optical or electrical systems** to interact with that sunlight in a deliberate way.

>> **Include a mechanism for delivering energy** to specific locations on Earth, rather than letting sunlight arrive naturally through the atmosphere.

The factor that varies and defines the different approaches to collecting and delivering energy is how the satellite interacts with sunlight and how the energy reaches the receiving technology on the ground. Some systems leave the sunlight largely unchanged by using mirrors to redirect it. Others convert sunlight into electricity in space and then transmit that energy wirelessly. These choices affect efficiency, safety, complexity, and cost, which is why space-based solar is better understood as a set of architectures rather than a single technology.

REMEMBER

Unlike Earth-based solar power systems, space-based solar is not constrained by local weather patterns or seasonal swings in the number of daylight hours. In theory, a satellite positioned correctly could provide illumination or power during early morning, evening, or even nighttime hours for a specific location. That potential is what makes space-based solar so compelling. The same features that make it powerful can make it technically demanding and expensive for now.

I want to separate modern space-based solar discussions from science fiction. The systems being developed are not intended to blanket the planet with constant energy or solve all grid challenges overnight. Instead, they involve targeted applications such as augmenting energy collection for existing solar farms, supporting solar power for regions that have more limited daylight hours, or providing energy during critical hours when traditional solar power production falls short.

Reflecting sunlight to Earth with orbital mirrors

The most intuitive form of space-based solar technology doesn't generate electricity in space at all. Instead, it relies on large orbital mirrors that reflect sunlight back toward specific locations on Earth. Think of these systems less as power plants and more as cosmic light managers that redirect sunlight to target areas that would otherwise miss out on collecting it.

In this approach, satellites equipped with highly reflective surfaces are placed in orbit to catch sunlight and reflect it toward the ground collectors at chosen times. The technology in space aims the reflected light at existing solar farms to extend their productive hours into early morning or evening, or at regions experiencing limited daylight. Because the sunlight remains sunlight all the way down to the Earth's surface, the ground infrastructure doesn't need to change dramatically. Conventional solar panels can capture the reflected light just as they would direct sunlight. Figure 16-1 shows an illustration of how a space-based solar power system that uses orbital mirrors might work.

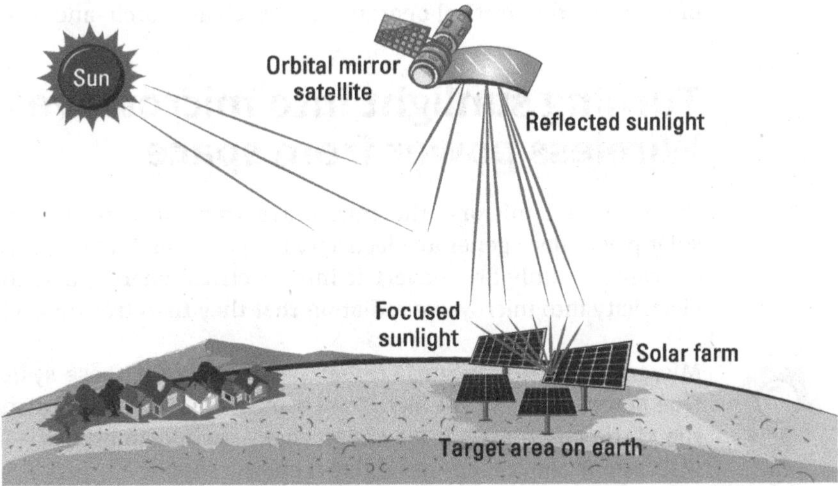

FIGURE 16-1: How space-based reflectors redirect solar energy to Earth.

Here are some advantages and challenges with space-based solar reflective technology:

>> **Simplicity is the mirror-approach's biggest advantage.** The technology needs no conversion of sunlight into electricity in space, no wireless power transmission, and no new receiving technology to install on the ground. Fewer conversion steps generally mean fewer losses of collected energy and fewer components that can fail.

>> **Reflecting sunlight from orbit is far from easy.** The mirrors must be large, lightweight, and precisely controlled. Even small alignment errors could scatter light away from the target or reduce its usefulness.

>> **Orbital mechanics add another layer of complexity.** Satellites in orbit move their positions constantly relative to the Earth's surface, which means reflected light windows are limited in duration unless multiple satellites work together. Coordinating these reflections to provide consistent, useful illumination requires sophisticated tracking and control systems.

>> **Practical and social considerations also exist.** Bright reflected light raises questions about potential glare, nighttime illumination that interferes with human sleep cycles and natural darkness, and impacts on astronomical studies or (especially nocturnal) wildlife.

Despite these challenges, orbital mirrors remain an attractive solution for space-based solar technology because they integrate cleanly with existing solar infrastructure. Rather than replace Earth-based solar power systems, orbital mirrors aim to stretch their usefulness by filling in the hours when the sunlight collection is low but energy demand is still high. They may also provide sunlight for the sake of lighting other critical operations — such as search-and-rescue missions.

Turning sunlight into microwaves: Wireless power from space

Unlike orbital mirrors, the microwave transmission approach to space-based solar power does generate electricity in space. Satellites equipped with solar panels capture sunlight, convert it into electrical energy, and then transform that electricity into microwave radiation that they then transmit wirelessly to Earth.

Microwave-based systems' researchers intend for these systems to function as true power plants in the sky. They want to deliver steady, dispatchable energy to the Earth-based grid. It's a vision that — while ambitious — fits naturally into discussions about long-term global energy supply.

Researchers have studied the basic concepts of this technology for decades. In orbit, solar panels operate without clouds, weather, or nighttime interruptions, which allows them to collect sunlight almost continuously. That electricity feeds microwave transmitters, which beam energy down to large receiving stations on the ground called *rectennas* (short for rectifying antennas). These rectennas convert the microwave signal back into electricity that can be fed into the grid. Figure 16-2 depicts this microwave transmission technology.

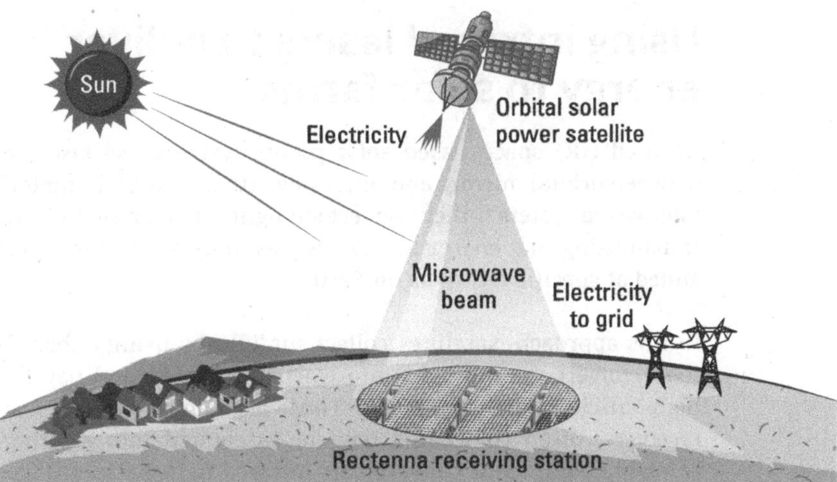

As with orbiting mirrors for collecting sunlight in space (see the previous section), space-based microwave technology has advantages and challenges. Microwave transmission of energy

>> **Has the appeal of predictability and controllability.** Microwaves can travel through clouds, rain, and dust with minimal loss, making them reliable regardless of weather. Engineers can also adjust the technology to spread the microwave beams over a wide area, keeping power density low enough to be safe for people, wildlife, and aircraft passing through.

>> **Introduces multiple conversion steps.** Each conversion — sunlight to electricity, electricity to microwaves, microwaves back to electricity — incurs losses in the collected energy. These losses mean that overall efficiency of microwave transmission technology is lower than that of a system that simply reflects sunlight. To compensate, microwave-based systems require very large solar arrays in orbit and equally large rectenna fields on Earth (often covering many square kilometers).

>> **Depends on infrastructure location on Earth.** Rectennas must reside in locations with ample open land, minimal interference concerns, and access to Earth-based transmission infrastructure. While rectennas don't need to be in remote deserts, they do require careful siting to balance land use, permitting, and community acceptance. (See Chapter 13 for some thoughts on local compatibility and acceptance.)

Researchers have successfully conducted demonstrations of wireless power transmission on Earth and in limited space experiments. The remaining hurdles are less about physics and more about economics, scale, launch costs, and international coordination.

Using infrared lasers to deliver energy to solar farms

Infrared (IR) space-based solar power systems use lasers and sit somewhere between orbital mirrors and microwave transmission in their characteristics. Like microwave systems, they convert sunlight into electricity in space. But instead of transmitting that energy as radio waves, they re-emit it as focused infrared light aimed at specific locations on Earth.

In this approach, satellites collect sunlight by using onboard solar panels; they then convert the sunlight into electrical energy and use the energy to power high-efficiency infrared lasers. Those lasers direct energy toward ground-based receivers, often co-located with or integrated into existing solar farms. The receiving equipment converts the infrared light back into electricity using specially designed photovoltaic cells tuned to the laser's wavelength. Figure 16-3 illustrates a space-based solar power system that uses lasers.

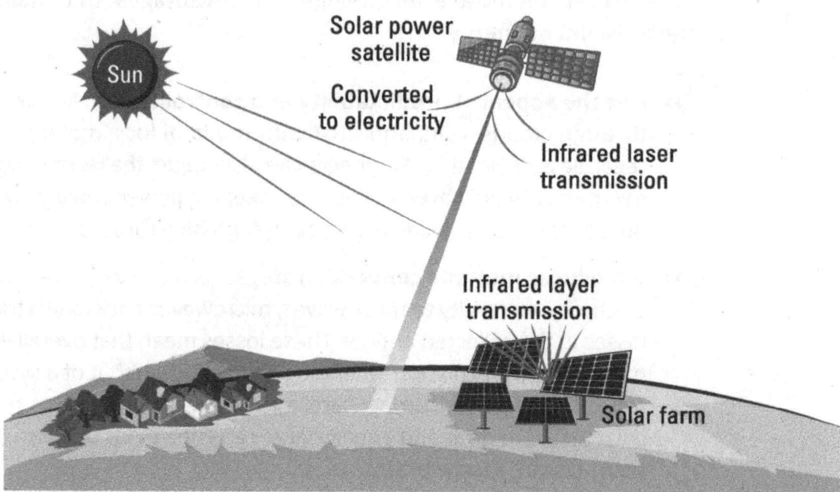

FIGURE 16-3: How infrared satellites work to collect and deliver solar power.

Here are advantages and challenges of using laser technology to deliver space-based solar power:

>> **Infrared lasers offer precision.** Laser beams can be tightly focused, allowing energy to be delivered to Earth-based infrastructure that has a much smaller footprint than microwave rectennas (see the preceding section). That makes laser-based systems particularly attractive for augmenting existing solar infrastructure, where land availability is limited and targeted delivery matters. Instead of building vast new fields for receiving solar energy from space,

lasers may be able to deliver energy directly to solar facilities that already have grid connections.

>> **Infrared lasers are more sensitive to atmospheric conditions than microwave technology.** Clouds, heavy moisture, dust, or smoke can scatter or absorb laser energy, which can reduce efficiency or interrupt energy delivery. This situation makes laser-based systems more weather-dependent and better suited to regions with consistently clear skies or to applications where intermittent delivery is acceptable.

>> **Overall efficiency is dampened by conversion steps.** The laser-based solar technology converts sunlight to electricity, electricity to laser light, and laser light back to electricity. Each step reduces total energy delivered, meaning more solar collection capacity is needed in orbit to achieve the required output on the ground. Thermal management aboard the satellite also becomes critical, as high-power lasers generate significant heat.

>> **Use of lasers requires safety rules and regulation.** Although space-based solar power systems that use lasers are designed to operate at power densities that avoid harm, the idea of directing laser energy from space raises understandable concerns. Any real-world deployment would require strict beam control, automatic shutoff systems, aviation coordination, and clear regulatory frameworks.

Despite the challenges — and like reflector solar satellites — infrared laser delivery of solar power from space to the Earth continues to attract interest because of its flexibility and compatibility with existing solar farms. Rather than trying to act as a standalone power source for entire regions, space-based solar systems that use lasers are often envisioned as a way to extend solar production into high-value hours, early mornings, evenings, or seasonal gaps, where Earth-based solar installations alone fall short.

Exposing the Impacts of Space-Based Systems on Solar Power

Space-based solar power matters not because it replaces ground-based solar, but because it changes the limits of when and where solar can contribute. Traditional solar is constrained by geography, weather, and the daily cycle of daylight. Space-based approaches challenge those constraints by making solar energy more flexible in its timing and, in some cases, more controllable.

None of the impacts discussed in this section are guaranteed. They depend on energy-related economics, regulation, public acceptance, and technical maturity. But when taken together, they explain why space-based solar power potential continues to attract serious attention — not as a silver bullet, but as a way to expand what solar power can do within an increasingly complex energy system.

Significant operational factors

One of the biggest potential impacts of space-based solar power on delivering needed power involves timing. On Earth, solar production peaks around midday, often when electricity demand can be relatively low. Demand usually rises in the early morning and evening, precisely when solar output drops. Space-based systems, especially mirrors and targeted laser delivery, often come up in discussion as tools for shifting solar availability into those higher-value hours.

This operational consideration occurs naturally because both mirror and laser space-based solar technologies utilize existing Earth-based solar farms. Rather than producing more energy overall, they could make existing solar generation more consistently useful to the grid.

Another operational impact involves power output *capacity*, a measure of how much energy a system produces compared to its theoretical maximum. Terrestrial solar plants typically operate at capacity factors of 20 to 30 percent, depending on their location and *overbuild* (intentionally installing more than the grid connection or inverter can handle so the system produces closer to its maximum more of the time). Space-based systems, operating above clouds and nighttime cycles, could achieve far higher utilization of their solar-collecting hardware. That doesn't automatically make them cheaper to operate (per MW produced), but it does change how solar power system planners think about reliability and predictability.

The value of solar power systems isn't just how much energy they produce, it's when that energy shows up.

Effects on Earth-based solar infrastructure

Space-based solar may also influence where Earth-based solar development happens. Regions that have limited available land, frequent cloud cover, or seasonal darkness often struggle to justify and right-size conventional solar installations. By delivering solar energy from orbit, space-based solar systems can help loosen constraints for these regional installations, potentially allowing solar to play a larger role in places where it currently underperforms.

At the same time, space-based solar power production introduces new kinds of grid interactions. Energy arriving via microwaves or lasers would likely be more easily controlled and delivered on demand than traditional solar energy, behaving less like a passive resource and more like a controllable generator. This capability could reduce reliance on *peaker plants* (power plants that operate only during periods of peak electricity demand) or energy storage in certain scenarios, which increases the need for careful coordination among grid operators.

Examining Barriers and Addressing Open Questions

If space-based solar power sounds promising but distant, that's because the biggest challenges aren't about whether it can work, they're about whether (and when) it can work at scale, safely, and affordably. Every approach described earlier in this chapter faces a mix of technical, economic, and social barriers that developers must address before space-based solar power production becomes just a regular piece of energy technology.

The barriers

The most obvious hurdle is cost, especially costs of launching space-based solar power projects. Even though rocket transportation seems to be cheaper and reusable launch systems more mature, placing large structures into orbit may still remain expensive. Space-based solar systems require large surface areas, whether mirrors, solar panels, or transmission hardware, and those structures must be lightweight, durable, and precisely controlled. Until launch and assembly costs drop, space-based solar will struggle to compete with terrestrial alternatives on price alone.

Another major challenge for space-based solar infrastructure involves complexity and reliability. Solar hardware in space must operate for years in a harsh environment with limited opportunities for repair. Systems that involve multiple energy conversions, active beam control, or coordinated *satellite constellations* (large groups of satellites working together in a network to provide coverage) raise the bar for reliability. Even small failures could interrupt energy delivery or reduce system performance, making system redundancy and fault tolerance essential — and expensive.

Space-based solar power generation doesn't need to match ground-based solar power on cost; there is a time value of energy, so it's competing with other ways of delivering power during critical hours.

The questions

Regulation and public acceptance are equally important for the success of space-based solar (or any solar) projects. But beaming energy from space — whether via microwaves or lasers — raises understandable questions about safety, coordination with airspace traffic, environmental effects, and international oversight. These systems would likely cross national boundaries, involve shared orbital space, and require new regulatory frameworks. Clear standards, transparency, and public trust would be prerequisites for deploying space-based solar power systems.

This new solar power category also raises questions about best-use cases. Should space-based solar aim to provide baseload power, peak power, emergency energy, or targeted augmentation for existing solar farms? Different answers to this question favor different system architectures, and it's still unclear which applications will justify the investment first. But early deployments are more likely to be specialized and incremental rather than global in scale.

The timing of space-based solar power development in conjunction with improving Earth-based systems reveals questions about coordination and collaboration. Advances in batteries, grid software, and terrestrial renewable technologies continue to reshape the energy landscape. Space-based solar doesn't exist in isolation; it must complement rapidly improving Earth-based solutions. Its role will depend as much on what happens elsewhere in the energy system as on its own technical progress.

Chapter **17**

Charging Batteries and Beyond: Storing the Sun

Solar power has one unavoidable limitation: It produces energy when the sun shines, but not necessarily when people want electricity. For years, that mismatch was treated as a drawback that limited how much solar power the grid could handle and make useful. Today, energy storage is changing the idea of useful solar energy. Instead of forcing the grid to adapt to solar power generation's schedule, storage allows solar energy to be captured, held, and released mostly on demand.

Storage lets you harness solar energy and move its distribution from the middle of the day to the evening — when demand rises and sunlight fades. But modern solar energy storage does far more than that. Batteries and other storage technologies help smooth out short-term fluctuations in power production, provide backup power, support grid frequency and voltage, and even replace traditional power plants during peak demand hours. In many regions, storage is no longer optional, it's becoming a core part of the design for new solar projects.

This chapter looks at storage not as a single technology, but as a toolkit. Different storage systems excel at different jobs. Some respond in milliseconds to stabilize the grid. Others discharge slowly over many hours. Some are compact and modular; others are massive and industrial. Understanding these differences helps explain why lithium-ion batteries dominate today's battery market, why alternatives such as flow batteries and gravity storage are gaining attention, and how storage choices shape the future of solar power.

Seeing Why Storage Matters for Solar Power

Solar power's goal and greatest strength — producing clean energy from sunlight — also presents its biggest challenge. The Sun doesn't follow or accommodate electricity demand but rises and sets on its own schedule. And clouds don't check in with grid operators before rolling through to block the Sun. As long as electricity-generating systems depended mostly on fuel-based power plants, the match between available power and demand was manageable. As solar penetration of the grid infrastructure increases, the mismatch between energy supply and demand becomes unavoidable.

Energy storage exists to solve the mismatch between supply and demand. It allows electricity generated at one moment to be used later, turning solar from a when-it-happens resource into a when-it's-needed one. Without storage, solar power generation must constantly adapt to the grid's needs. With storage, the grid can adapt to solar.

REMEMBER

Seeing why storage matters means understanding that it isn't a backup plan or an accessory. It's an enabling technology that's essential to the long-term growth of solar power. Without storage, solar remains constrained by nature. With storage, solar becomes something closer to a dispatchable resource; one that can be planned, scheduled, and relied upon.

Increasing the usefulness of solar power production

Early solar systems leaned heavily on the existing power generation and delivery system to provide flexibility. Gas plants ramped up when the sun went down. Transmission lines moved excess power elsewhere. Demand responded to changing quantities of available energy imperfectly. Those tools (power plants and transmission lines) still matter, but storage adds a new layer of control that didn't exist before.

Adding storage to a solar power system doesn't make more solar energy, it makes solar energy more useful.

Storage also changes how the grid values solar energy generation. Instead of being paid only for the energy it produces at noon, solar paired with storage can deliver power during evening peaks, respond instantly to grid disturbances, and even defer expensive grid upgrades. In many markets, these additional capabilities and accessibility are now worth as much as, or more than, the energy itself.

Enabling solar power at scale

Just as importantly as adding useful value, storage allows solar power production to scale. Grids can tolerate only so much uncontrolled power generation from midday sunlight before system reliability suffers. Storage smooths that variability, absorbs excess production, and releases it later in a controlled way. That's why regions with high solar adoption increasingly require or incentivize storage as a part of or add-on to new solar power projects.

This controlled use of solar power through storage isn't limited to utility-scale systems. Homes, businesses, microgrids, and remote communities all benefit from storage for different reasons: resilience, cost savings, reliability, and/or grid independence. The storage technology in various use cases may look similar, but the value it provides depends on context.

Shifting supply to when it's needed

One of the most straightforward roles of energy storage is shifting the delivery of solar power from when it's generated to when it's actually needed. Solar production tends to peak in the middle of the day when sunlight is strongest and panels are most productive. Electricity demand, however, often rises later, in the early evening, when people return home, turn on lights and appliances, and run heating or air conditioning. Storage bridges that timing gap.

By storing excess solar energy produced during sunny hours and releasing it later, batteries allow solar power to compete directly with traditional *peaker plants* (traditional, usually fossil-fuel-based power plants) that are built to run only during periods of high demand. Stored solar electricity that's delivered during these peak hours is usually more expensive and more valuable to the grid, which means storage increases the economic impact of each kilowatt-hour of solar energy.

Figure 17-1 illustrates this relationship between solar energy production and use. Solar production peaks around midday, right when time-of-use electricity rates are often lowest. Without storage, much of that energy would either be exported to the grid at a low value or curtailed entirely. With battery storage capability, that midday surplus is captured and saved and then discharged later as prices rise and demand increases.

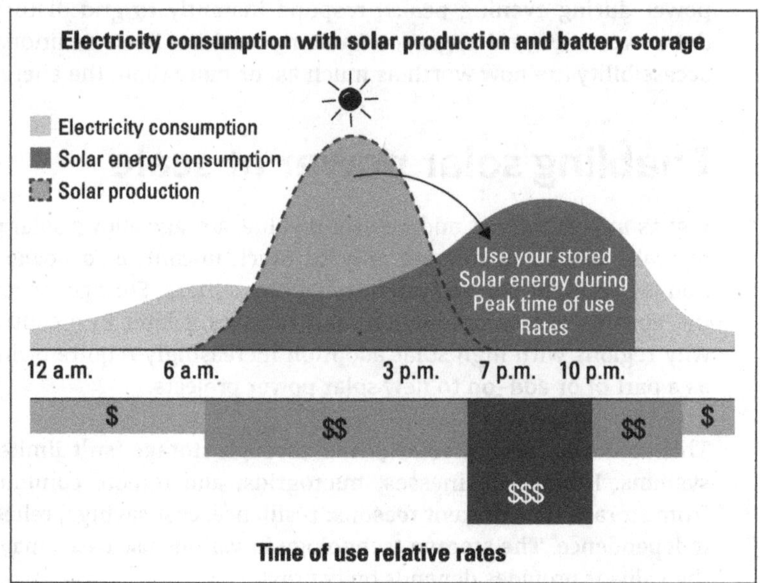

FIGURE 17-1: Solar production versus energy rates with storage.

Time-shifting of solar power delivery is important because it:

>> **Facilitates lower-cost energy use.** Later in the day, as the sun sets and electricity prices climb, the battery storage system can supply solar energy for use instead of drawing power from the grid. The result isn't more solar power generation, but better-timed solar power use, which is why storage is so effective under time-of-use rate structures (find more information about utility rate structures in Chapter 2).

>> **Reduces strain on the grid.** Without storage, large amounts of midday solar can overwhelm local grid infrastructure, forcing operators to curtail production or move power inefficiently across long distances. Storage absorbs the surplus energy and releases it later, making better use of both generation and transmission assets.

>> **Provides flexible power release.** Some storage systems are designed for short bursts of power lasting minutes, while others discharge steadily for several hours. The appropriate duration depends on the energy need being addressed — whether that need involves evening-hour peaks, overnight coverage, or longer power disruptions.

As power system planners grow more confident that solar energy can be stored and dispatched when needed, storage increasingly becomes a standard companion to new solar installations rather than an optional upgrade.

Smoothing intermittency and grid services

Even when solar energy is delivered at the right time of day, it still may not be perfectly steady. Clouds pass over, weather systems move in, and solar output can rise or fall faster than traditional power plants are able to respond. As solar energy supply penetration increases, these short-term fluctuations become more visible and more challenging for grid operators.

Fluctuation in solar energy production offers storage another role that's different from simple time-shifting of energy delivery (see the preceding section). Batteries respond almost instantly when called upon. They can absorb excess power when solar output spikes and inject power when output dips, thereby smoothing out the power fluctuation on time scales ranging from seconds to minutes. To the grid, solar power production paired with storage behaves less like a variable resource and more like a controlled one.

Time-shifting changes the period when solar energy is used; smoothing intermittency controls how steadily the energy is stored and delivered.

Beyond smoothing out power fluctuations, storage systems provide a range of grid services that help maintain reliability. Solar energy storage systems can:

>> Help with *frequency regulation* (keeping the grid's electrical frequency within tight bounds) and other grid requirements, such as voltage support, spinning reserve (backup generation that's already running and ready to increase output immediately), and fast-response capacity during power disturbances. Most batteries excel at these tasks because they don't need to ramp up; they respond immediately.

>> **Replace functions formerly handled by large fossil-fueled power generators.** Instead of keeping gas plants running just in case they're needed, grid operators can rely on batteries to respond in milliseconds and then recharge when conditions stabilize. This improves efficiency while reducing fuel use and emissions.

>> **Help protect grid infrastructure.** By smoothing rapid changes in power flows, batteries reduce stress on transformers, substations, and transmission lines. Over time, this can delay or eliminate the need for expensive grid upgrades, which is one reason why utilities often support storage even when energy prices alone don't justify it.

In many electricity markets, these grid services are explicitly valued. Storage systems may earn revenue not just by shifting energy, but by participating in ancillary service markets that reward speed, accuracy, and reliability. This shifts storage from a passive companion to solar power systems into an active grid asset that allows more solar power to be added to the grid without compromising system stability and keeping supply and demand balanced.

Exploring Types of Batteries

After you understand why storage matters, the next question is what kind of storage actually does the job. Not all batteries are built the same, and the differences matter far more than brand names or buzzwords. Some batteries respond instantly but don't maintain power delivery for over time. Others deliver power slowly for hours. Some are compact and modular; others are massive and industrial.

To make sense of storage options, it helps to step back from individual technologies and focus on a handful of core characteristics that determine how a battery behaves in the real world. These traits shape everything from system design to economics to grid value.

There is no "best" battery to use with solar power systems, only batteries that are better suited to specific jobs.

REMEMBER

Four characteristics show up again and again when engineers, utilities, and developers evaluate storage systems. Together, these characteristics describe how much energy a battery can store, how long it can deliver that energy, how controllable delivery is, and how much energy is lost along the way. They are:

>> **Capacity.** Measured in kilowatt-hours (kWh), capacity tells you how much energy a battery can hold.

>> **Duration.** A battery's duration describes how long the battery can deliver the energy it contains at a given power level. A 4-hour battery behaves very differently from a 12-hour one.

>> **Dispatchability.** The characteristic of dispatchability refers to how quickly and precisely a battery can respond to commands, which matters enormously for grid services.

>> **Round-trip efficiency.** A battery's round-trip efficiency measures how much energy you get back compared to the energy you put in.

These characteristics explain why lithium–ion batteries dominate today's market: They're compact, efficient, and highly responsive. But they also explain why alternatives such as flow batteries, sodium–based batteries, and even gravity storage continue to attract interest. Those technologies trade speed and density for longer duration, simpler materials, or improved safety at scale. Table 17-1 shows you how battery's core characteristics line up across different storage technologies.

TABLE 17-1   **Comparison of Storage Technologies**

Storage Technology	Typical Capacity	Typical Duration	Dispatchability	Round-Trip Efficiency (RTE)
Lithium-ion (Li-ion)	Low to medium	1–6 hours	Very fast (milliseconds)	85–95%
Flow batteries (iron, vanadium)	Medium to high	1–12+ hours	Fast (seconds)	65–85%
Sodium-based batteries	Medium	2–8 hours	Fast (seconds)	75–90%
Gravity storage	Very high	4–24+ hours	Slow to moderate	70–95%
Green hydrogen	Extremely high	Days to seasons	Slow (minutes to hours)	24–45%
Compressed air energy storage (CAES)	Very high	8–100+ hours	Moderate	45–70%
Thermal (heat) batteries	High	Hours to days	Slow	70–95%

Lithium-ion battery technology

Lithium–ion batteries are the default choice for most modern solar–plus–storage systems, and for good reason. They strike a rare balance between high efficiency, fast response, and compact size. Whether installed in a home garage, behind a commercial building, or alongside a utility–scale solar plant, lithium–ion systems have become the benchmark against which other storage technologies are measured.

Strengths	Trade-offs
High-efficiency, compact, excellent controllability	Higher upfront cost and limited long-duration capability

What sets lithium-ion apart is how well it performs across the characteristics that matter most for today's solar systems. These batteries store a meaningful amount of energy in a relatively small footprint, deliver that energy quickly and precisely, and lose very little energy in the process. That combination makes them especially well-suited for daily cycling — charging during sunny hours and discharging during evening peaks — as well as for grid services that demand rapid response.

Lithium-ion battery storage systems paired with solar energy collection work well as

>> **Providers of short- to medium-duration storage,** typically ranging from one to six hours. This aligns closely with the most common use cases for storage today: shifting solar energy into higher-value evening hours, reducing demand charges, and smoothing short-term variability on the grid. Within this window, lithium-ion delivers excellent performance and strong economics.

>> **Active grid assets,** by responding almost instantly to control signals, allowing them to act not just as passive energy storage, but as active grid assets. They can inject power, absorb excess generation, or stabilize grid frequency in fractions of a second. From a grid operator's perspective, that speed and precision are extremely valuable and difficult for slower storage technologies to replicate.

>> **Cost-effective and reliable storage,** developed through decades of investment in lithium-ion technology — driven by consumer electronics and electric vehicles — that created a mature manufacturing ecosystem. That scale has pushed costs down, improved performance, and standardized safety and control systems in ways that newer storage technologies are still working toward.

Lithium-ion's strengths also define its limits. These batteries are optimized to be used frequently, not to sit idle for long stretches. Storing energy for weeks or months without regular cycling accelerates degradation and makes poor economic use of the system. Extending lithium-ion storage to multi-day or seasonal durations also requires adding large numbers of battery cells, which quickly drives up cost, complexity, and material requirements.

In short, lithium-ion excels at fast, efficient, everyday storage, but it is not designed for long-term energy banking. That gap is exactly where other storage technologies begin to make sense, not as replacements for lithium-ion, but as complements that solve problems lithium-ion wasn't built to address.

Iron flow battery technology

Iron flow batteries take a very different approach to energy storage than lithium-ion. Instead of packing energy into solid cells, flow batteries store energy in liquid electrolytes held in external tanks. The battery itself is really two parts: a power component that determines how quickly energy can be delivered and an energy component (the tanks) that determines how much can be stored. This separation gives flow batteries a unique advantage when longer durations of energy delivery are needed.

Strengths	Trade-offs
Long duration and long cycle life	Larger footprint and lower efficiency than lithium-ion

Iron flow batteries have these advantages:

» **Scalability.** Because capacity is tied to tank size rather than battery cells, iron flow batteries scale more gracefully for longer-duration storage, typically four hours and beyond. Want more energy? You add bigger tanks. Want more power? You upgrade the stack. That flexibility makes flow batteries especially attractive for utility-scale solar projects in which space is available and long discharge windows matter more than compact size.

» **Frequency of use.** Iron flow batteries also handle frequent cycling well. They can be charged and discharged deeply without the same degradation concerns seen in lithium-ion systems. For applications that involve steady, predictable cycling day after day, this durability can translate into long service life and stable performance over time.

» **Availability.** Iron flow batteries also benefit from relatively simple and abundant materials. Compared to some other battery chemistries, they avoid many supply-chain constraints and thermal risks, which can be appealing for large installations designed to operate for decades.

The trade-off comes in efficiency and footprint. Flow batteries generally lose more energy in the storage process than lithium-ion and require significantly more physical space. Their response time is fast enough for many grid applications, but not instantaneous, which makes them less competitive for high-value, fast-response grid services. As a result, they tend to be used where duration and longevity matter more than speed.

In practice, iron flow batteries aren't trying to replace lithium-ion. They fill a different niche. Where lithium-ion excels at short bursts of high-performance storage, iron flow batteries step in when systems need to deliver energy steadily for many hours without wearing themselves out.

Sodium-based battery technology

Sodium-based batteries are one of the newer entrants in the energy storage landscape, but they're gaining attention quickly because they promise many of the benefits of lithium-ion without some of its constraints. Instead of relying on lithium, these batteries use sodium, an element that is far more abundant and widely distributed. That abundance makes sodium-based storage appealing from a long-term cost and supply-chain perspective.

Strengths	Trade-offs
Abundant materials and improving safety profile	Still emerging with less field history

In terms of performance, sodium-based batteries tend to land somewhere between lithium-ion and flow batteries. They offer solid efficiency, relatively fast response times, and moderate energy density. While they don't yet match lithium-ion's compactness or market maturity, they perform well enough across capacity, duration, and dispatchability to serve many of the same use cases, particularly at commercial and grid scales.

Attractive features of sodium-based storage include:

>> **Safety.** Many sodium chemistries are more thermally stable than lithium-ion, reducing fire risk and easing concerns about siting. That can be especially important for large installations near buildings or critical infrastructure. Lower safety overhead also has the potential to reduce system costs as the technology matures.

>> **Duration.** Many designs of sodium-based batteries target multi-hour storage in the two-to-eight-hour range, making them suitable for solar energy shifting, peak shaving, and load balancing. While they're not designed for seasonal storage, they compete well in the same daily and intra-day window in which lithium-ion currently dominates.

The biggest limitation today is commercial readiness. Sodium-based batteries are still scaling into widespread deployment, and long-term performance data is more limited than for lithium-ion. Costs are falling, but manufacturing volumes remain smaller, which means fewer suppliers and less standardization.

That said, sodium-based batteries are often viewed as a bridge technology: not a radical departure from electrochemical storage, but a refinement that trades extreme energy density for material availability, safety, and potentially lower long-term costs.

Gravity storage technology

Gravity storage takes one of the simplest ideas in physics and applies it at grid scale: Lift something up when energy is plentiful and let it fall when energy is needed. Instead of storing energy chemically, gravity-based systems store energy mechanically by raising heavy masses and then converting that potential energy back into electricity as the masses descend.

Strengths	Trade-offs
Extremely long life with minimal degradation	Requires large physical structures and specific sites

You can implement this concept to construct a gravity storage system in several ways. Some systems use pumped hydro, moving water uphill into a reservoir and releasing it through turbines later. Others rely on solid masses, such as concrete blocks or steel weights, lifted by cranes or winches and lowered when power is required. While the designs vary, the underlying principle is the same.

Here are a couple of good points for gravity storage:

>> **It shines in capacity and longevity.** These systems can store very large amounts of energy for many hours or even days, and they don't degrade the way batteries do. Gravity-based storage has no chemical reactions to wear out, which means that these systems can operate for decades with relatively little performance loss.

>> **Efficiency is respectable (but not exceptional).** Energy losses occur during lifting, friction, and power conversion, which means gravity storage typically returns less energy than lithium-ion batteries over short cycles. However, for long-duration applications in which durability and scale matter more than speed, those losses may be acceptable.

The trade-offs are speed and siting. Gravity storage systems generally respond more slowly than electrochemical batteries and require specific physical conditions for a site, such as elevation changes, open land, or large structures. These restrictions limit where developers can build these systems and makes them impractical for small or distributed applications.

In practice, gravity storage isn't competing with lithium-ion or sodium-based batteries for everyday solar-plus-storage systems. Instead, it fills a niche in which long duration, massive capacity, and long asset life outweigh the need for compactness or fast response.

Green hydrogen technology

Green hydrogen storage takes solar energy and converts it into a fuel rather than storing it directly as electricity. When excess solar power is available, electricity is used through a process called electrolysis to split water into hydrogen and oxygen. The hydrogen is then stored and later converted back into usable energy, either by burning it in a turbine or using it in a fuel cell.

Strengths	Trade-offs
Enables season storage and fuel flexibility	Very low efficiency for electricity-to-electricity use

Strengths and advantages of green hydrogen include:

>> **Scale and duration.** Unlike batteries, hydrogen can store enormous amounts of energy for very long periods of time, ranging from days to entire seasons. This makes green hydrogen one of the few technologies capable of addressing the challenge of long-term energy storage, such as saving summer solar energy for use in winter.

>> **Hydrogen's flexibility.** Once produced, it isn't limited to electricity generation. It can be used as an industrial feedstock, blended into gas networks, or used as fuel for transportation. That versatility allows hydrogen systems to serve multiple roles within a broader energy system, not just solar storage.

But the trade–offs of green hydrogen technology are significant and involve

>> **Far lower efficiency.** Battery-based storage is much more efficient when used for electricity-to-electricity applications. Large amounts of energy are lost during electrolysis, compression or liquefaction, storage, and conversion back to electricity. As a result, hydrogen is rarely the most economical option for daily energy shifting or short-duration storage.

>> **Complexity and high capital-investment.** Electrolyzers, storage vessels, safety systems, and conversion equipment add layers of infrastructure that aren't required for battery systems. Because of this, hydrogen storage is most attractive at large scales, where long-duration resilience or fuel flexibility justifies the added complexity and cost.

In practice, green hydrogen isn't a competitor to lithium–ion or flow batteries. It addresses a completely different problem: how to store renewable energy when time horizons stretch from hours to months.

Compressed air technology

Compressed air energy storage (often called CAES) stores energy by using electricity to compress air and store it in a sealed space, such as an underground cavern, pipeline, or large pressure vessel. When energy is needed, the compressed air is released, heated if necessary, and expanded through a turbine to generate electricity.

Strengths	Trade-offs
Scales to massive systems	Depends on suitable geology and has moderate efficiency losses

CAES is best known for its ability to deliver very large-scale, long-duration storage. These systems can store energy for many hours or even days, making them well-suited for regional grid balancing and renewable integration at scale. Because air itself is abundant and inexpensive, the limiting factor isn't materials but geology and infrastructure.

One of the main advantages of compressed air is its ability to scale economically where conditions allow. Underground caverns, such as salt domes or depleted gas fields, can hold vast amounts of energy without the need for large quantities of battery materials. That makes CAES attractive in regions with suitable geology and high renewable penetration.

The trade-offs for using CAES systems involve:

>> **Lower efficiency.** Traditional compressed air systems lose a significant portion of energy during compression, storage, and expansion, especially when heat generated during compression isn't fully recovered. Newer designs aim to improve efficiency by capturing and reusing that heat, but CAES still trails electrochemical batteries in round-trip performance.

>> **Slower response times.** While compressed air technology can ramp power up and down reliably, they aren't built for instantaneous response or fine-grained grid services like battery storage systems are. As a result, they tend to complement batteries rather than replace them, handling bulk energy shifting while faster technologies manage short-term variability.

In practice, compressed air storage occupies a middle ground between batteries and hydrogen. It offers much longer duration than lithium-ion, better efficiency than hydrogen, and lower material intensity than many battery systems, but only where the physical conditions make it feasible.

Thermal battery technology

Thermal batteries store energy not as electricity, but as heat. Instead of charging electrochemical cells, these systems use electricity to heat a material, such as molten salts, ceramic bricks, sand, or other high-heat media, and then release that stored heat later when it's needed. In some cases, the heat is used directly; in others, it's converted back into electricity.

Strengths	Trade-offs
Very low cost and durable for heat storage	Limited usefulness for electricity delivery

The advantages of using thermal energy storage include:

>> **Simplicity and cost.** Heat is far easier and cheaper to store than electricity, especially at large scales. Thermal batteries rely on abundant materials, avoid complex chemistry, and can operate for many years with minimal degradation. For applications that need heat rather than power, their efficiency can be extremely high.

>> **Storage for longer durations.** Thermal systems often store energy for hours or days with relatively low losses. This makes these storage systems particularly useful for industrial processes, district heating, and buildings that require consistent thermal energy. In those settings, thermal batteries can replace fossil fuels directly, without ever converting energy back into electricity.

The main limitation of thermal energy storage is flexibility. Thermal batteries are highly effective when heat is the end goal, but less efficient when electricity is required. Converting stored heat back into electrical power introduces additional losses, which makes thermal storage less competitive for general-purpose grid applications.

Because of this, thermal batteries tend to be task-specific rather than universal solutions. They don't compete head-to-head with lithium-ion batteries for fast response or with hydrogen for seasonal storage. Instead, they carve out a niche where low cost, long life, and direct heat delivery matter most.

Chapter **18**

The Future of Solar Cells

Solar panels may look simple, but the technology inside them is anything but static. Behind the scenes, scientists, engineers, and manufacturers are constantly refining solar cells for composition, efficiency of operation, and ultimately, appropriate disposal. This chapter looks ahead at where solar technology is going, not in a sci-fi way, but in ways that are already shaping today's market.

Some of the changes in solar power system construction and operation are about solving practical problems, like recycling old panels and reducing material waste. Others focus on squeezing more electricity out of the same amount of sunlight, which can lower costs and expand where solar makes sense. You'll also hear a lot about new materials and cell designs — such as perovskites and tandem cells — that promise to improve performance beyond what traditional solar panels can achieve on their own.

This chapter ties innovations back to a trend you've seen throughout the book: falling costs. Even as solar panels become more advanced, they continue to get cheaper to produce and install. Understanding why that's happening, and how long it's likely to continue, helps put the future of solar in perspective.

Recycling Solar Panels

Solar panels are designed to last a long time. Most come with warranties of 25 to 30 years, and many continue producing electricity well beyond that timeframe. But long lifespans don't eliminate an important question: What happens to solar panels when they finally reach the end of their useful life?

Solar panels consist mostly of familiar materials. The largest share of their bulk is glass, followed by aluminum framing, plastics, copper wiring, and semiconductor materials such as silicon. The good news is that many of these materials can be recycled by using existing industrial processes. The challenge lies not in whether panels can be recycled, but in how to do so efficiently, economically, and at scale.

Recycling now and looking ahead

Early solar installations are only now beginning to retire in significant numbers, which means recycling infrastructure is still catching up. In many regions, panels are currently processed through specialized facilities that disassemble them, recover the aluminum frames and glass, and extract valuable metals. As volumes of used-up solar panels grow, recycling must become more automated and cost-effective, much like the process that happened with electronics and automotive batteries.

Solar panels don't become waste overnight. Recycling is a long-term challenge that's growing alongside the success enjoyed by the increase in solar power installations.

Recycling may not grab headlines the way new solar cell technologies do, but it's a critical part of the solar power industry's maturation. As the industry evolves, managing materials responsibly ensures that solar remains not just a clean energy source, but a sustainable one from start to finish.

Recycling that's sustainable and built-in

Recycling also plays an important role in sustainability. Recovering materials reduces the need for new mining, lowers the environmental footprint of future panels, and helps address concerns about waste. From a lifecycle perspective, solar panels already produce far more energy than is required to manufacture them. Recycling pushes that balance even further in solar energy's favor.

One reason recycling hasn't been a major bottleneck so far is timing. Solar waste volumes remain small compared to other waste streams, and many panels are still performing well decades after installation. But as solar deployment accelerates worldwide, planning for end-of-life management becomes increasingly important.

Policy and design choices are starting to reflect that reality. Some manufacturers are designing panels with easier disassembly in mind, while regulators in certain regions require recycling plans or extended producer responsibility programs. These efforts aim to make recycling a built-in feature of solar's future, not an afterthought.

Maximizing Panel Efficiency

When people talk about *better* solar panels, they often mean one thing: efficiency. In simple terms, efficiency describes how much of the sunlight hitting a panel converts into usable electricity. Higher efficiency means you get more power from the same amount of space, and that matters whether panels are going onto a rooftop or parking structure, or into a utility-scale solar farm.

Most commercial silicon solar panels today convert roughly 20 to 23 percent of incoming sunlight into electricity. That number may not sound impressive at first, but it represents decades of steady improvement. Early solar cells struggled to reach double-digit efficiency, and each incremental gain since then is the result of increasingly sophisticated engineering.

Components of efficiency

Efficiency maximization is about squeezing more value out of every ray of sunlight. While breakthroughs tend to arrive gradually rather than overnight, each small improvement in efficiency compounds across millions of panels, which pushes solar power systems' performance and affordability steadily forward.

Here are components and considerations of solar power efficiency:

>> **The amount of space devoted to the system:** Efficiency matters most where space is limited or expensive. On a small roof, a higher-efficiency panel can produce noticeably more electricity than a lower-efficiency one, even if both occupy the same footprint. In large projects, higher efficiency can reduce the number of panels, racking, and wiring needed, which can lower overall system costs even if the panels themselves are more expensive.

>> **Overall engineering of systems:** Improving efficiency isn't just about better materials; it's also about reducing losses. Engineers work to minimize electrical resistance inside the cell, prevent light from reflecting away before it's absorbed, and manage heat buildup that can reduce performance. Many modern panels use textured surfaces, advanced coatings, and improved internal wiring layouts to capture more sunlight and move electricity more efficiently.

>> **Real-life limitations:** A practical limit exists for how far traditional system designs can increase efficiency. Single-junction silicon cells are approaching their theoretical efficiency ceiling, which means future gains from silicon alone are likely to be smaller and harder won. That reality is one reason researchers are exploring new materials and multi-layer designs. I talk about these topics in the next sections of this chapter.

Perovskite solar cells

Perovskite solar cells are one of the most talked-about developments in solar technology and for good reason. In just over a decade, they've gone from laboratory curiosity to record-setting performers, achieving efficiency gains faster than any solar material before them. While traditional silicon took decades to approach its current limits, perovskites have closed much of that gap in a remarkably short time.

The term *perovskite* doesn't refer to a single material, but to a class of crystalline structures that are especially good at absorbing light and moving electric charge. In a solar cell, the perovskite layer sits between charge-transport layers, as shown in Figure 18-1, where sunlight excites electrons that are then guided toward electrodes to produce electricity. The layered design looks complex, but the layering is actually one of perovskite's biggest strengths: Each layer can be tuned independently to improve performance.

Noting advantages

One reason perovskites generate so much excitement is how efficiently they absorb sunlight. Very thin perovskite layers can capture a broad range of the solar spectrum, which means that thinner solar panel material can do the same job as thicker silicon wafers. This advantage opens the door to lighter panels, flexible designs, and lower manufacturing costs — at least in theory.

The perovskite cell manufacturing process is another major advantage because the process requires relatively low-temperatures, which. involves printing and coating techniques more similar to making newspapers than melting silicon in giant furnaces. If manufacturers can scale those processes reliably, perovskites could dramatically reduce the energy and cost required to produce solar cells.

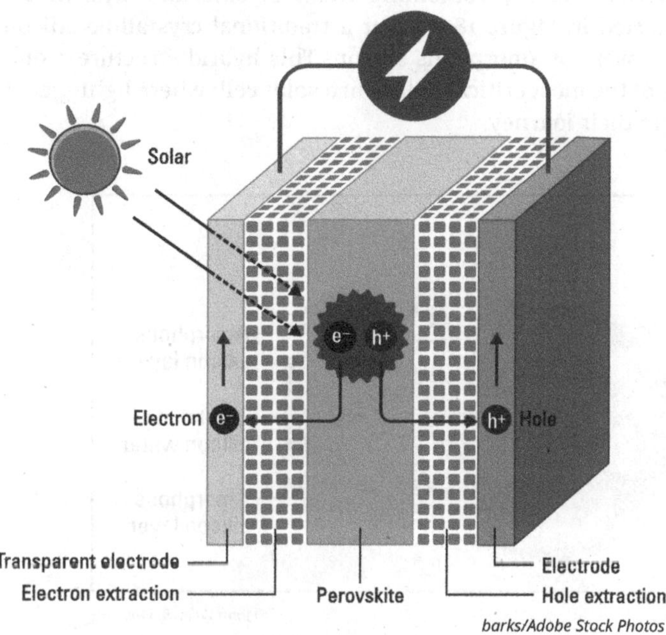

FIGURE 18-1: Perovskite solar cell architecture.

barks/Adobe Stock Photos

Recognizing the downsides

Despite the promise, perovskites aren't ready to replace silicon in rooftop solar panels just yet. The biggest challenge is durability. Early perovskite cells degraded quickly when exposed to moisture, heat, or ultraviolet light — conditions solar panels must survive for decades. Although perovskite panel stability has improved significantly in recent years, long-term performance in real-world environments remains the key hurdle to their widespread adoption.

Another downside issue is materials management. Some high-performing perovskite formulations contain small amounts of lead, which raises questions about environmental handling and recycling. Researchers are actively working on lead-free alternatives and robust encapsulation methods, but manufacturers must address these concerns before solar power systems can deploy perovskites at scale.

For now, perovskites are best understood as a bridge technology, one that could either stand alone in future panels or, more likely, work alongside silicon to push solar performance beyond today's limits.

Heterojunction solar cells

Heterojunction (HJT) solar cells improve performance not by reinventing solar materials, but by combining them in smarter ways. At their core, HJT cells (depicted in Figure 18-2) pair a traditional crystalline silicon wafer with ultra-thin layers of amorphous silicon. This hybrid structure reduces energy losses at one of the most critical points in a solar cell: where light-generated electrons first begin their journey.

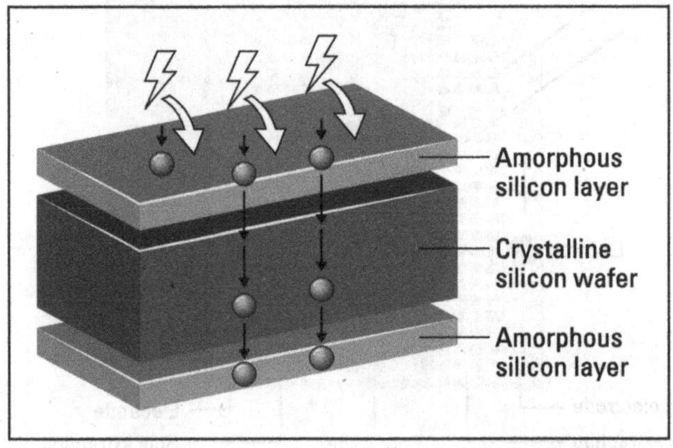

FIGURE 18-2: HJT cell architecture.

© John Wiley & Sons, Inc.

In a conventional silicon solar cell, microscopic defects at the surface of the silicon wafer act like tiny potholes, trapping electrons before they can be collected as usable electricity. HJT cells address this problem through *passivation*, a process in which thin layers of amorphous silicon coat the crystalline silicon surface. The coating smooths out the defects in the silicon wafer and allows electrons to move more freely. The result is higher efficiency without significantly increasing cell thickness or complexity.

Rather than replacing silicon, HJT refines it. That's why many industry observers see heterojunction technology as one of the most practical near-term paths to higher efficiency before more radical designs, such as tandem cells (see the section "Tandem cells" later in the chapter), become widespread.

Here are some advantages and trade-offs related to HJT technology:

>> **Standout temperature performance.** Solar panels become less efficient as they heat up, which matters in real-world installations in which panels often operate well above laboratory climate conditions. HJT cells lose less efficiency at high temperatures than many conventional silicon designs, making them particularly attractive in warm climates and utility-scale installations in which thermal losses add up quickly.

>> **High bifacial performance.** Because the cell structure is symmetric and well-passivated (smoothed out) on both sides, HJT panels can efficiently collect reflected light from the ground or surrounding surfaces, as well as from the Sun. This trait makes them well suited for installations over concrete, sand, snow, or reflective ground cover, where *backside generation* (collecting of reflected light) meaningfully boosts output.

>> **Manufacturing complexity.** HJT cells require more precise fabrication steps and additional materials compared to standard silicon cells. That situation has slowed adoption somewhat, but as manufacturing scales and processes improve, HJT is increasingly viewed as a bridge technology. HJT panels are more advanced than today's mainstream panels but compatible with existing silicon supply chains.

Tunnel oxide passivated contact

Tunnel oxide passivated contact (TOPCon) solar cells take aim at one of the most stubborn sources of energy loss in silicon photovoltaics: what happens when electrons reach the edges of a solar cell and have nowhere productive to go. Even in high-quality silicon, electrons can recombine at contacts instead of flowing into the external circuit, quietly stealing efficiency. TOPCon solar cells are designed to stop that recombining from happening.

The name that underlies the TOPCon acronym (tunnel oxide passivated contact) sounds intimidating, but it describes a surprisingly elegant idea. A TOPCon cell adds an ultra-thin insulating layer — so thin that electrons can quantum tunnel through it — between the silicon wafer and the electrical contact. (*Quantum tunneling* means that the electrons can pass through a barrier that has a higher kinetic energy and appear on the other side as if passing through a wall.) This tunnel oxide layer (see Figure 18-3) blocks unwanted recombination while still allowing charge carriers to pass along efficiently. In effect, it acts like a selective gate: Electrons get through; power losses do not.

This contact design dramatically improves voltage and overall cell efficiency without changing the fundamental silicon absorber. That's why TOPCon is often described as a "drop-in" advancement. It builds on existing crystalline silicon technology while squeezing more usable electricity out of every photon that enters the cell.

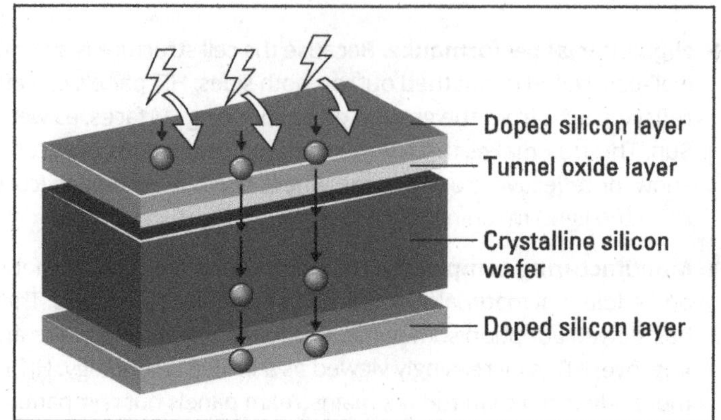

FIGURE 18-3: TOPCon cell architecture.

TOPCon technology also has these advantages:

>> **TOPCon cells perform well under real-world conditions.** They show strong results at partial shading, maintain efficiency at higher temperatures, and degrade more slowly over time than earlier silicon architectures. These traits make them appealing not just in lab settings, but in utility-scale projects in which long-term performance and reliability matter as much as headline efficiency numbers.

>> **TOPCon cell manufacturing generally uses more conventional manufacturing equipment and materials** when compared to heterojunction cells. This advantage has helped accelerate adoption, and many manufacturers see TOPCon as a practical next step that improves performance without requiring a complete overhaul of silicon production lines. That balance — meaningful efficiency gains with manageable complexity — is why TOPCon has moved quickly from research labs into commercial deployment.

Like HJT, TOPCon cell design doesn't replace silicon; it refines how silicon works. And like HJT, it also serves as a stepping stone. The same contact innovations used in TOPCon cells are now being incorporated into more advanced designs, including tandem cells (see the next section), in which minimizing every possible loss becomes even more important.

Tandem cells

Figure 18-4 shows the basic idea behind tandem solar cells. Instead of relying on a single material to do all the work, tandem cells stack multiple solar materials on top of one another so each layer captures a different portion of sunlight.

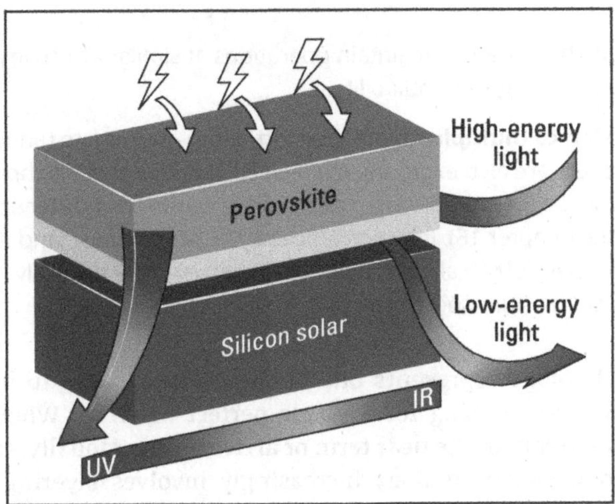

FIGURE 18-4: Tandem cell architecture.

© John Wiley & Sons, Inc.

Traditional solar cells face a fundamental trade-off. A single material can be very good at absorbing certain wavelengths of light, but it inevitably wastes others — mainly as heat. Tandem cells solve this problem by dividing the solar spectrum among layers. The top cell layer absorbs high-energy light while lower-energy light passes through to layers below, where it can still convert into electricity.

The most widely studied tandem solar cell design today pairs perovskite on top of silicon. Silicon serves as a reliable bottom layer that absorbs lower-energy light, while the perovskite layer captures higher-energy photons that silicon handles less efficiently. Together, the two materials outperform either one alone and push efficiencies beyond the practical limits of *single-junction* (with one layer of semiconductor material) silicon cells.

Here are some intriguing aspects of tandem cell engineering:

>> **Building on technology that already exists:** Tandem cells are especially compelling because they build on existing technologies rather than replacing them outright. In many designs, the silicon bottom cell is similar to those in the solar panels already manufactured at scale, with the perovskite layer added on top. That compatibility could allow tandem cells to piggyback on today's manufacturing infrastructure while delivering a meaningful jump in performance.

>> **Early positive performance results.** Tandem cells are already breaking efficiency records in laboratory settings, and pilot programs for manufacturing at scale are underway. The biggest challenges remaining are durability and complexity. Stacking multiple materials involves more interfaces, more sensitivity to moisture and heat, and tighter quality control requirements.

Ensuring that all layers maintain operations at similar effective rates is essential for long-term reliability.

>> **Expansion to multiple layers.** Beyond silicon–perovskite tandem cells, researchers are also exploring *multi-junction cells* that stack three or more materials. These multiple-layer designs are common in outer space applications (see Chapter 16), in which efficiency outweighs cost, and they point toward what's physically possible when materials are carefully matched to maximize the solar spectrum.

Tandem technology represents one of the clearest paths to higher solar panel efficiency without waiting for a single perfect material. Whether tandem cells become mainstream in the near term or arrive more gradually, they show how the future of solar cell technology increasingly involves layering existing, proven layers and trying out new combinations — not just inventing new solar cell materials from scratch.

Decreasing Costs

A counterintuitive aspect of solar technology is that it keeps getting cheaper even as it becomes more advanced. In many industries, better performance can mean higher prices. Solar power technology has flipped that pattern. Over time, improvements in materials, efficiency, and design have consistently pushed adoption costs down rather than up.

The result is a reinforcing cycle. As solar power adoption becomes cheaper, people deployed it more widely. As deployment grows, manufacturing improves and refines. As manufacturing improves, costs fall again. That cyclical dynamic helps explain why solar power production has moved from being a niche technology to one of the most cost-competitive sources of electricity in the world.

Solar power's future isn't just about better technology; it's about better overall economics driven by scale, efficiency, and innovation.

Here are aspects of the decreasing cost of deploying solar power systems:

>> **Economies of scale.** As solar deployment has grown worldwide, manufacturers have learned how to produce cells and modules more efficiently, with less waste and tighter quality control. Equipment has improved; supply chains have matured; and competition has driven innovation. These effects compound, meaning that each new generation of technology benefits from everything that came before it.

>> **Efficiency improvements.** When a solar cell converts more sunlight into electricity, solar power systems need fewer panels to produce the same amount of energy. That reduces not just module costs, but also racking, wiring, labor, land use, and maintenance. Even modest gains at the cell level can translate into meaningful savings across an entire system.

>> **New cell architectures.** Advances in solar cell technology — such as HJT, TOPCon, and tandem designs discussed in the section "Maximizing Panel Efficiency" earlier in the chapter — don't just aim for higher efficiency. They also aim for better value per watt of power production. Some technologies cost more per panel but reduce overall system costs. Others focus on durability and longevity, spreading costs over more years of energy production. What matters isn't the price of a single panel, but the total cost of energy production over the panel's lifetime.

>> **Recycling and materials innovation.** Recovering aluminum, glass, and semiconductor materials reduces reliance on new raw materials and stabilizes supply chains. Meanwhile, research into alternative materials and thinner cells aims to reduce the amount of material used without sacrificing performance.

Chapter **19**

Making Solar Smart with Software

Solar panels may quietly turn sunlight into electricity, but software is what makes modern solar power systems truly intelligent. From the earliest design sketches to day-to-day operation years later, digital tools guide decisions, catch problems early, and help systems deliver more value than hardware alone ever could.

This chapter looks at how software fits into the solar power story at every stage. You find out how design platforms model sunlight, predict production, and optimize system layouts before a single panel is installed. You discover how monitoring and operations software track performance in real time, flagging issues that might otherwise go unnoticed for months or years.

Increasingly, software tools are being enhanced via artificial intelligence (AI). In the solar world, AI isn't about replacing engineers or installers; it's about handling complexity. Machine-learning models can sift through huge amounts of weather data, performance history, and sensor readings to improve energy output forecasts, detect subtle performance issues, and recommend adjustments more quickly than a human ever could.

Also in this chapter, you get to zoom out to the bigger picture of energy management. As solar power generation pairs with batteries, electric vehicles, and smart appliances, software becomes the brain that coordinates the interconnection. AI-driven energy management systems can learn usage patterns, respond to time-of-use rates, and automatically decide when to store energy, when to use it, and when to export it to the grid.

The power of software for support and management transforms solar power from a passive energy source into an adaptive, responsive part of the modern energy system.

Designing and Optimizing the Best System

Before a single panel is installed, solar power systems live their first life in software. Design and optimization tools help installers, engineers, and developers answer a deceptively simple question: What's the best way to turn sunlight at this location into usable electricity? The answer depends on far more than just roof size or panel count.

Design and optimization software turn sunlight into numbers long before it becomes electricity. By the time panels go up, the software helps make most of the important decisions — and tests, refines, and optimizes them in the digital world.

Combining digital tools for system design

Modern solar design software starts with the basics: location, orientation to the path of the Sun, and space available for installation. Using satellite imagery, light detection and ranging (LiDAR) data, and site measurements, these tools map out the roofs, ground areas, and obstructions in three dimensions. That mapping enables designers to model how sunlight moves across a site throughout the year and identify where panels will perform best.

From there, optimization becomes a balancing act. Designers configure panel placement, tilt, spacing, and electrical layout to maximize energy production while staying within real-world constraints like building codes, fire setbacks, structural limits, and budgets. The goal isn't simply to achieve the highest theoretical power output; it's to create the best-performing system that can be built economically. Increasingly, as solar installations have become cheaper, designers optimize for overall system performance and cost-effectiveness rather than squeezing the maximum output from each individual panel. In other words, when the hardware is inexpensive, efficiency gains elsewhere matter more.

Weather and climate data can play a major role as well. Design software pulls in historical solar *irradiance data* (measures of the Sun's intensity), temperature data, and cloud-cover data to estimate how much electricity a system can produce month by month and year by year. These projections form the backbone of financial models, which help owners understand expected savings and payback timelines.

Pulling in historical data for optimization

Increasingly, AI is enhancing the software involved in the solar power system design process. Machine-learning models can analyze thousands of prior solar installations to spot patterns humans might miss. For example, AI can suggest layout adjustments that slightly reduce shading losses, flag designs that historically underperform in certain climates, or refine production forecasts based on subtle local weather trends. Instead of replacing designers, AI acts like a fast-learning assistant that improves the accuracy of appropriate system configuration and output projections over time.

Optimizing a solar power system extends beyond the panels themselves. Design software models inverter sizing, *string configurations* (how the panels connect together), and potential electrical losses to ensure energy flows efficiently from panel to meter. For solar systems paired with batteries, optimization tools evaluate how storage should be charged and discharged under varying rate structures and usage patterns.

Operations & Maintenance Software

After a solar power system is built and turned on, software doesn't disappear. In many ways, it becomes even more important. Operations and maintenance (O&M) software is what keeps solar systems performing as expected over years and decades, not just on day one. O&M software turns a solar installation from a "set it and forget it" asset into a well-managed power plant. It's the quiet layer of intelligence that keeps systems reliable, bankable, and producing value year after year.

At its core, O&M software answers a simple question: Is the system doing what it's supposed to be doing right now? Monitoring platforms track energy production in real time and compare actual output against expected performance. If a system underperforms, the software flags it quickly, often before an owner would ever notice a problem on their electric bill.

Modern monitoring tools don't just show total energy produced; they have other functions and benefits:

>> **Detailing component performance specifics:** They break down performance by inverter, string, or even individual panel, depending on system design. That level of detail makes it much easier to pinpoint issues such as shading from new obstructions, inverter faults, wiring problems, or panel degradation. Instead of sending a technician to inspect an entire site, operators can target exactly where data indicates that something is wrong.

>> **Assigning problem urgency:** As solar fleets grow larger, especially for commercial and utility-scale systems, O&M software becomes essential for prioritization. Not every performance issue noted deserves the same level of urgency. Software helps operators rank problems based on lost energy, safety risk, and repair cost, ensuring that the most impactful fixes happen first.

>> **Detecting subtle performance anomalies:** AI is increasingly part of the ongoing monitoring process. Machine-learning models can detect subtle performance anomalies that don't trigger simple alarms. For example, AI might recognize that a particular inverter is slowly degrading compared to similar units at other sites, even if that inverter hasn't technically failed. That foresight enables predictive maintenance: fixing or replacing equipment before it causes significant downtime.

>> **Supporting long-term asset management:** O&M platforms track system performance over time, document maintenance history, and help owners verify warranty claims. For investors and lenders, this data provides confidence that a system is being properly cared for and will continue generating revenue or saving you money as expected.

Consistent monitoring matters more than occasional check-ins. Small problems caught early are far cheaper to fix immediately than large problems discovered months later.

Energy Management Systems and Smart Home Integration

If O&M software focuses on keeping solar systems healthy, energy management systems (EMS) focus on making them *smart*. EMS platforms sit one layer above monitoring and control how energy flows through a home or building in real time. Instead of just showing you what's happening in a system, they help decide what should happen next. As solar power system adoption grows, software increasingly determines how valuable a system can be, not just how large it is. Panels capture energy, but software decides how intelligently that energy is used. In that sense, making solar power smart may be one of the most important steps in the technology's evolution.

Simplifying solar power with EMS

At a basic level, an EMS coordinates solar production, battery storage, and electricity consumption. It decides when to store energy, when to use it on-site, and when to export it to the grid. In homes and small commercial buildings, this choice often means prioritizing *self-consumption* — using solar power when it's available instead of buying electricity from the utility.

Energy management systems also change how people experience solar power usage. Instead of thinking about panels, inverters, or kilowatt-hours, users interact with dashboards, apps, and automation settings. The technology fades into the background, and the system feels more like a service than a piece of infrastructure.

Using EMS at home or for an office

Smart home integration takes EMS management a step further by connecting solar power systems to everyday devices. Thermostats, water heaters, EV chargers, pool pumps, and major appliances can all be scheduled for power distribution or controlled based on solar production, electricity prices, or grid signals. For example, a smart system might pre-cool a home in the afternoon by using solar power, which reduces the need for grid electricity during expensive evening hours.

This whole-house or office coordination becomes especially powerful under time-of-use (TOU) rates. When electricity prices change throughout the day, an EMS can automatically shift some usage loads to cheaper periods without the homeowner or small-business operator having to think about it. Charging an electric vehicle, running laundry, heating water, or cooling the space during office hours becomes a background optimization problem handled by software rather than habits.

Creating advantages for the grid

For utilities and grid operators, EMS-enabled homes and commercial buildings represent a new kind of flexibility. When thousands of systems respond intelligently to grid conditions, they can collectively reduce peak demand, absorb excess solar power generation, and provide grid services that formerly required large, centralized power plants. This network is sometimes referred to as a *virtual power plant,* even though it's built from many small systems acting together.

AI is also increasingly central to these systems. Instead of relying on fixed schedules or simple rules, AI-driven EMS platforms learn from past behavior, weather forecasts, and utility price signals. Over time, they get better at predicting when energy will be produced and when it will be needed. That predictive capability allows EMS platforms to make smarter decisions, such as reserving battery capacity for an upcoming heat wave or adjusting charging behavior ahead of a known price spike.

The Part of Tens

Get a practical checklist of smart steps to take before going solar — from understanding your energy use to lining up permits, incentives, and future needs.

Discover the biggest benefits of solar power, including lower bills, cleaner energy, greater independence, and long-term value for homes and businesses.

Clear up common myths and misunderstandings about solar power so you can separate outdated assumptions from how modern solar power actually works.

Chapter **20**

Ten Smart Moves before Adopting Solar

This chapter is your solar pre-flight checklist. Before committing to panels, contracts, or financing, a handful of smart moves can save you money, reduce surprises, and lead to a better-performing system. These steps don't require you to become a technical expert; they simply help you ask the right questions at the right time.

Think of this chapter as a way to slow down just enough to make good decisions. A little preparation upfront may mean a smaller system, lower costs, smoother permitting, and fewer (or no) regrets later. If you work through the ten items in this chapter, you'll be in a much stronger position to move forward with your solar power system installation confidently.

Get a Firm Grasp on Your Current Energy Usage

Before you go forward with a solar power system installation, it's essential to understand how much electricity your home actually uses and how that usage is distributed over time. Solar systems are designed and priced around your consumption, so having a clear baseline helps ensure you get a system that fits your needs and budget. Chapter 2 covers more information about evaluating your energy use.

Start by collecting at least 12 months of electric bills. Look for your total annual usage in kilowatt-hours (kWh), which smooths out seasonal changes like summer air conditioning or winter heating. If your utility offers online usage data, review it to see daily or hourly patterns, which can reveal whether most of your electricity use happens during the day or in the evening.

The timing of your energy use matters because many utilities now use time-of-use rates, in which electricity costs more during peak hours. If your highest usage coincides with expensive periods, a solar power system alone may not offset your biggest energy costs if you don't include storage capabilities or *load shifting* (changing your peak usage to times when the grid sees less demand).

When evaluating your energy use, also consider which energy needs may change. Working from home, adding an electric vehicle, or switching appliances from gas to electric can significantly increase future demand. Because solar power systems are long-term investments, it's worth sizing them with tomorrow's energy needs in mind.

Improve Efficiency First to Reduce System Size and Cost

Before investing in solar panels, look for ways to reduce how much energy your home needs in the first place. Improving efficiency can lower your electricity use, which often means installing a smaller solar power system, lower upfront cost, and faster payback. See Chapter 5 for more information about finding ways to reduce your home's energy needs.

Start by making simple upgrades that deliver immediate savings. Replacing older light bulbs with LEDs, sealing air leaks around doors and windows, and upgrading inefficient appliances can noticeably reduce electricity consumption without major renovations. These improvements are usually far cheaper than adding extra solar panels that you may not need after you make your home more efficient.

Heating and cooling deserve special attention, since they often account for the largest share of household energy use. Better insulation, sealed ductwork, and smart thermostats can dramatically cut demand. In many cases, efficiency upgrades also qualify for rebates or tax incentives, making them even more cost-effective.

By reducing waste first, you make every solar panel more valuable. Solar power systems work best when paired with homes that use energy wisely.

Understand Your Utility Rate Structure

Not all electricity is priced the same, and understanding how your utility charges for power can make a big difference in how valuable a solar power system can be for you. Two homes with the same annual energy use can see very different savings depending on their utility's rate structure.

Start by identifying which type of rate plan you're on. Many utilities now use *time-of-use* (TOU) rates, in which electricity costs more during peak hours (often evenings) and less during off-peak times. Others use flat rates or tiered pricing, in which costs increase as usage rises. Your electric bill or utility website should clearly state which plan applies to your usage.

Your billing plan matters because of timing. Solar power systems produce the most energy during the middle of the day. If your electricity is most expensive later in the evening, solar power alone may not offset your highest-cost usage unless you pair it with energy storage or smart load shifting (as mentioned in the section "Get a Firm Grasp on Your Current Energy Usage" earlier in the chapter).

Understanding your utility's rate structure also helps you evaluate potential future changes in that structure. Utilities are increasingly moving toward time-based pricing, so even if you're on a flat rate today, your savings picture may change over time.

Identify Tax Credits and Rebates

Incentives can significantly reduce the cost of installing and using a solar power system, but only if you know what incentives are available and how to claim them. Federal, state, local, and utility programs all play a role, and the details can vary widely depending on where you live. Chapter 7 offers a good look at incentives and subsidies.

Start your investigation with federal incentives, which are often the largest. Tax credits reduce the amount of income tax you owe. Although the federal government has eliminated tax credits for home installations as of this writing, some related commercial enterprises still have access to them, which may translate into a savings passed down to a home installation. Beyond that, many states, cities, and utilities offer rebates, performance incentives, or exemptions that lower upfront or ongoing costs.

Timing matters for qualifying for and claiming incentives for a solar power system. Some incentives have limited funding or change over time, and eligibility can depend on your system size, installation date, or whether you own the system outright. Always verify current rules before signing a contract for a solar power system.

Contractors can help identify incentives, but it's still worth doing a bit of homework yourself. Knowing what programs exist makes it easier to compare bids and spot assumptions that may not apply to your situation.

Get Multiple Bids from Vetted Contractors

One of the easiest ways to avoid overpaying or ending up with the wrong system is to get more than one solar power system quote. Labor pricing, equipment choices, and system designs can vary significantly between contractors, even for a system planned for the same home.

Aim to get at least three bids from reputable installers. Make sure each proposal clearly shows the system size, estimated production, equipment models, warranties, and total cost. Having all the necessary information in each bid makes comparing bids easier; you do not need to guess if you're comparing apples to apples.

Pay attention to how contractors explain their recommendations. A good installer should be able to justify system size, panel placement, and assumptions about energy use and incentives. Be cautious of bids that seem rushed, overly aggressive, or dramatically cheaper without a clear explanation.

Vet the solar vendor companies themselves. Look for proper licensing, insurance, experience in your area, and recent customer reviews. Solar power systems are long-term investments, and the installer you choose matters just as much as the equipment.

Assess Site Readiness and Conditions

Not every home is immediately ready for a solar power system installation, and understanding your site conditions upfront can prevent delays, redesigns, or unexpected costs. A strong site candidate doesn't have to be a perfect one, but it does need to meet some basic criteria:

>> **Start with your roof or available space.** Roof age, condition, and roofing materials matter since panels are typically installed for decades. If your roof needs replacement soon, it's often smarter to handle that first. Also consider shading from trees, neighboring buildings, chimneys, or vents. Even partial shading of the roof area where the solar panels go can significantly reduce output.

>> **Orientation and structure matter, too.** South-facing roofs (in the Northern Hemisphere) usually perform best, but east- and west-facing surfaces can still work well depending on your energy use and rates. Your roof also needs to be structurally sound enough to support the added weight of panels and racking.

Don't forget practical details such as electrical panel capacity, attic access, and available space for inverters or batteries. These factors don't usually make or break a project, but they can affect cost and complexity.

Forecast Future Energy Needs

Solar systems last 25 years or more, so it's important to think about how your energy use might change over that time. A system sized perfectly for today can feel undersized just a few years down the line if your household's electricity needs grow, as mentioned in the section "Get a Firm Grasp on Your Current Energy Usage" earlier in the chapter.

Common changes include adding an electric vehicle, switching from gas to electric appliances, installing heat pumps, or expanding your home. Even lifestyle changes, like working from home more often, can significantly increase daytime electricity use.

REMEMBER

You don't need to predict future energy needs perfectly. Instead, identify likely upgrades and discuss them with your installer so they can design a system that allows for expansion or accounts for higher expected demand. In many cases, slightly oversizing upfront or choosing expandable equipment is cheaper than retrofitting later.

Know How You Need to Connect to the Grid

Not all solar power systems connect to the grid in the same way, and understanding your options upfront can save time, money, and frustration later. Most homes install grid-tied systems, which stay connected to the utility and draw power when the solar system isn't producing enough. Others add battery backup to store excess energy for later use, which changes how the system is designed, permitted, and operated.

Your utility's interconnection rules matter here. Some require specific inverter types, system size limits, or additional studies before approving a connection. If you're considering batteries, backup power for outages, or future off-grid capability, those choices affect wiring, equipment, and costs from day one.

You don't need to master the technical details, but you should know whether you want a simple grid-tied system or something more complicated and expandable. That clarity helps your installer design a system that fits both your goals and your utility's requirements.

Make Sure to Map Permits, Approvals, and Inspections

Solar installation projects rarely fail because of technology, but instead, they stall because of paperwork. Permits, utility approvals, and inspections all sit on the critical path between signing a solar power installation contract and flipping the switch to energize the system. Knowing what approvals are required, and in what order, helps you avoid long delays. See Chapter 6 for more information.

Before Permission to Operate (often called PTO) is granted, most residential solar projects need at least three layers of approval: a local building permit, utility interconnection approval, and a final inspection. Some jurisdictions add design reviews, fire department sign-offs, or homeowner association approvals on top of that. Each approval step can take weeks to achieve and can interrupt installation progress if you don't plan for it.

A good, full-service installer can manage this approval process for you, but you should still ask how long each step typically takes and what dependencies exist. If inspections are backed up or utility approvals move slowly in your area, the timeframe in which your system actually starts producing power will be impacted.

Figure Your Payback Horizon and Risks

A solar power system almost always saves money over time, but how long you must wait to pay off your investment and what factors may change along the way matter just as much as the final savings number. Your payback horizon is the number of years required for your energy savings to equal what you initially spent on the system. Chapter 4 offers a closer look at payback from a solar power installation.

Payback depends on a handful of variables: system cost, available incentives, electricity rates, and how much of your solar power you actually use. In high-rate areas, payback can be surprisingly fast. In lower-rate markets, savings still add up, just more gradually.

You should also understand the risks involved that affect the payback period. Changes to *net metering rules* (whether, when, and how utilities credit solar energy system owners for excess electricity they export to the grid), utility rates, or tax incentives can affect the actual cost savings. Equipment performance may slowly degrade over time, and future electricity usage may rise or fall depending on lifestyle changes. None of these considerations invalidate having a solar power system, they simply mean that payback projections should be realistic, not overly optimistic.

The goal in understanding your investment in solar power isn't to chase the shortest possible payback at all costs. It's to understand the range of possible outcomes so that you're confident in the decision to go forward with a solar power installation even if conditions change.

Chapter **21**

Ten of the Best Benefits of Going Solar

Solar power isn't just about producing electricity differently, but it changes how you think about energy altogether. Whether you're a homeowner, a business owner, or a community leader, adopting solar power delivers a mix of financial, practical, and environmental benefits that few other investments can match.

This chapter walks through ten of the biggest advantages of going solar. Some show up immediately as cost savings on your electric bill. Others build quietly over time, improving property resilience, energy cost stability, and long-term home value. Together, the advantages explain why solar adoption continues to accelerate across homes, businesses, and entire power systems.

Lower Electricity Costs

For most people, the first and most tangible benefit of solar is simple: lower electricity costs. When you generate your own power, you buy less from your utility, and every kilowatt-hour you don't purchase is money saved.

A solar power system can effectively replace a portion of your electricity bill after a one-time investment in installation. Instead of paying a utility company month after month at rates you don't control, you help lock in the cost of energy production upfront. Over time, that trade almost always works in your favor, especially as utility rates continue to rise.

Even if adding a solar power system doesn't eliminate your electric bill entirely, reducing the bill can have a meaningful impact on household budgets or business operating expenses. For businesses, those savings improve margins and free up capital for other priorities. For homeowners, they make monthly budgets more predictable and manageable.

Energy Price Stability

One often underrated benefit of having a solar power system is price stability. Utility rates for electricity tend to rise over time due to fuel costs, infrastructure upgrades, regulatory changes, and growing demand. When you rely entirely on grid power, you're exposed to all of those forces that can raise rates, whether you like it or not.

Having a solar power system changes that situation. After your system is installed, the cost of producing electricity from your panels is essentially fixed. You have no fuel to buy and no surprise rate hikes tied to global markets or utility investments. Each year when utility prices increase, the value of your solar-generated electricity quietly grows.

This stability is especially valuable for businesses and long-term homeowners. Predictable energy costs make budgeting easier and reduce risk, turning electricity from an uncertain operating expense into a far more manageable budget item.

Tax Incentives and Rebates

Solar power generation has moved so quickly from niche installations to mainstream adoption because governments at multiple levels have consistently supported this transition with financial incentives. The incentives are designed to lower the upfront cost of installing a solar power system and shorten the time it takes for the system to pay for itself. In practice, that means many solar power system buyers don't actually pay the full sticker price of their system; instead, they take advantage of upfront or ongoing subsidies.

The most common incentive is a tax credit, which reduces the amount of income tax you owe, dollar for dollar. Unlike a deduction, which only lowers your taxable income, a credit directly lowers your tax bill. State programs may offer these incentives, whereas federal tax credits for individuals adopting solar power disappeared in 2025. Local programs may also offer rebates, performance-based incentives, or exemptions from sales or property taxes. These types of cost savings vary widely by location, but together they can shave a meaningful percentage off the total cost of a system.

What makes incentives especially powerful is how they stack with other benefits. Lower upfront costs improve payback, reduce financing needs, and make solar accessible to a much wider range of households and businesses. Even when incentives gradually change or phase down, they tend to remain part of the solar power industry landscape because policymakers view solar investments as infrastructure, not as a luxury purchase.

Carbon Emissions Reduction

Solar power's environmental benefit is straightforward and powerful: It produces electricity without emitting carbon dioxide during operation. Every kilowatt-hour of energy generated by a solar panel is one that doesn't need to be produced by using a fossil-fuel power plant somewhere on the grid. Over time, replacing fossil-fuel energy with green energy adds up to a meaningful reduction in greenhouse gas emissions, even for relatively small systems.

For many households and businesses, installing a solar power system is the single largest step they can take to lower their carbon footprint. Unlike changes in behavior that rely on constant effort, solar reductions happen automatically, every time the sun shines. After you install a solar power system, the emissions savings continue quietly in the background for decades.

The impact of using a solar power system is cumulative. As more systems come online, the grid itself becomes *cleaner* (relying less on fossil fuels) and reduces emissions for everyone, not just solar power system owners. That's why utilities and regulators track solar capacity so closely: Each new installation nudges the overall power mix away from fuel-based generation and toward cleaner sources.

Boosted Property and Asset Value

Solar panels don't just lower energy bills; they can also increase the value of the property on which they're installed. Buyers increasingly recognize solar power systems as an income-producing asset rather than a cosmetic upgrade. A home or building that has a solar power system comes with lower operating costs baked in, which makes the property more attractive in markets with energy prices that are high or rising.

Multiple studies have shown that solar-equipped homes often sell more quickly and at a premium compared to similar properties without a solar power system. The logic is simple: Future buyers inherit the benefits of reduced electricity costs without having to pay the upfront installation expense themselves. For commercial properties, a solar installation can improve net operating income (NOI), which directly influences valuation in income-based appraisals.

Having available solar power can also protect property value over time. As utilities raise rates and buildings without solar become more expensive to operate, properties with on-site power generation can stand out as more resilient and cost-efficient. This situation is especially true for long-term owners who view their property as a durable asset rather than a short-term investment.

Energy Independence

Adopting solar power generation gives you more control over where your electricity comes from and how much you rely on outside providers. Instead of depending entirely on a utility for power, a solar system lets you produce at least part of your electricity right where you use it. That shift doesn't mean you're completely off the grid, but it does mean you're less exposed to outages, price hikes, and supply disruptions.

For homeowners and businesses alike, energy independence is about resilience. When paired with battery storage, solar power can keep critical energy demands satisfied during grid outages, whether those outages come from storms, heat waves, wildfires, or equipment failures. Even without batteries, generating your own power during the day reduces how much energy you need to pull from the grid overall.

You can also receive a psychological benefit that many solar power system owners don't anticipate. Producing your own energy changes how you think about energy consumption. When you can see power being generated in real time, energy becomes tangible rather than abstract. That awareness often leads to smarter usage, further increasing the value of the system.

Improvements in ESG Scores and Public Perception

Producing solar power has become a visible, credible way for organizations to demonstrate responsibility in the Environmental, Social, and Governance (ESG) arenas. By generating clean electricity on-site or sourcing it through solar projects, companies can directly reduce their carbon footprint and report measurable emissions reductions. That kind of transparency matters more than ever as customers, investors, and regulators increasingly expect climate-friendly commitments to be backed by real action.

From an ESG standpoint, solar touches all three pillars.

>> **Environmentally,** it lowers emissions and local air pollution.

>> **Socially,** solar projects often support community development, local jobs, and energy access.

>> **On the governance side,** investing in long-term clean energy signals thoughtful risk management and future-oriented leadership.

Together, these factors can improve ESG ratings used by investors, lenders, and partners to assess corporate performance.

Public perception is another powerful, and sometimes underestimated, benefit. A visible solar installation on a roof, parking canopy, or campus sends a clear message about values. For businesses, that message can strengthen brand trust, attract sustainability-minded customers, and help recruit employees who care about working for responsible organizations. Even for homeowners, a solar power installation can signal environmental awareness and forward-thinking choices to neighbors and future buyers.

As sustainability reporting becomes more standardized, solar provides something especially valuable: hard data. Kilowatt-hours produced, emissions avoided, and costs stabilized are all metrics that can be tracked and communicated clearly.

In a world in which *greenwashing* (the sometimes deceptive portrayal of a company, product, or policy as environmentally friendly) is increasingly scrutinized, solar stands out as a concrete, verifiable step toward cleaner energy.

Job Creation and Local Economic Growth

Solar power doesn't just produce electricity; it creates jobs across the economy. Every project supports work in manufacturing, system design, installation, inspection, operations, and maintenance. Because solar systems must be installed and serviced locally, much of this employment stays in the community rather than being outsourced, which supports local businesses and skilled trades.

At the residential and commercial level, solar power installations drive demand for electricians, engineers, installers, and permitting professionals. Utility-scale projects amplify this impact, creating hundreds of construction jobs and long-term operations roles, often in rural areas where new economic opportunities can be scarce. These projects also generate secondary economic activity for suppliers, transportation services, restaurants, and lodging.

Solar power investments can strengthen local economies over the long term. They generate lease payments for landowners, tax revenue for municipalities, and stable employment tied to infrastructure that lasts decades. As solar power system adoption grows, it also encourages workforce training and clean-energy career pathways, helping communities build expertise in an industry that is expanding rather than declining.

Silent and Low-Maintenance Operation

One of the most underrated benefits of solar power is how quietly it does its job. After installation, solar panels generate electricity without engines, moving parts, or combustion. There's no noise, vibration, or exhaust, just steady power production whenever sunlight is available. This feature makes solar power generation especially well-suited for homes, offices, schools, hospitals, and dense urban areas where noise and disruption matter.

Solar power systems are also remarkably low maintenance compared to most other energy technologies. Panels are designed to withstand decades of exposure to weather, including rain, wind, heat, and snow. In most cases, maintenance amounts to little more than periodic inspections and occasional cleaning to remove dust or debris. Inverters and monitoring systems may need replacement or servicing over time, but these events are infrequent and predictable.

Because solar power systems have so few mechanical components, operating costs stay low and reliability stays high. There's no fuel to manage, no oil to change, and no complex mechanical wear cycle to plan around. For homeowners and businesses alike, this simplicity translates into fewer surprises, lower long-term costs, and peace of mind knowing the system will quietly keep working in the background year after year.

Modularity and Scalability

The scope of a solar power system can easily grow with you. Solar systems are inherently modular, meaning they're built from individual panels that you can add, remove, or rearrange over time. You don't have to get the system configuration perfect on day one. You can often expand a system sized for today's needs later when electricity use increases, budgets change, or technology improves.

This design flexibility is especially valuable as lifestyles and energy demands evolve. A homeowner may start with a modest system and later add panels for charging an electric vehicle, running a heat pump, or heating and cooling a home addition. Businesses can scale solar power installations alongside facility expansions or rising operational loads. In many cases, system owners can layer in battery storage later, turning a basic solar installation into a more resilient and sophisticated energy system.

Scalability also reduces risk. Instead of committing to a single, all-or-nothing investment, a solar power installation allows for incremental decision-making. You can test performance, understand savings, and build confidence before expanding further. That adaptability is a major reason why solar power systems work across such a wide range of homes and commercial projects.

Chapter **22**

Ten Misconceptions about Solar Power

Solar power is one of those topics for which nearly everyone has an opinion, but many of those opinions are based on how solar power systems worked decades ago and not how they work today. Early solar systems were expensive, inefficient, and limited in where they could be used. Some of the myths from that era stuck around long after the technology moved on.

In this chapter, I help you tackle the most persistent misconceptions head-on. These ideas aren't trick questions or edge cases; they're the kinds of concepts that people bring up at dinner parties, town halls, and budget meetings. The chapter gives you a clearer picture of what solar power systems actually do well, where limits still exist, and why these systems have become a mainstream part of the energy system rather than a niche experiment.

Solar Power Only Works When the Sun Is Shining

This misconception about reliability comes from thinking about solar power as a real-time power source instead of an energy system that works over time. It's true that solar panels generate electricity from sunlight, but that doesn't mean they suddenly become useless when clouds roll in or the sun goes down. You find that

>> **Solar panels produce power whenever there is daylight,** not just during bright, cloudless conditions. Diffuse sunlight on overcast days still generates electricity, and cooler temperatures can actually improve panel efficiency. That's why places with moderate climates often perform just as well annually as places that are typically sunny.

>> **Most solar systems are connected to the grid.** During the day, panels generate electricity that either powers the home directly or flows out to the grid. At night or during low-production periods, electricity simply flows back from the grid, just as it always has. From the user's perspective, power availability feels continuous, and solar quietly reduces how much grid electricity you need overall.

>> **Competitive and scalable storage changes the picture even further.** Batteries as part of a solar power system allow you to store excess solar energy produced during the day and use it later. This capability smooths the gap between production and use. Solar power isn't about shining as the primary energy resource all the time; it's about producing enough energy across days, seasons, and years to meaningfully offset conventional power.

You Need to Live in a Hot, Sunny Climate for Solar to Be Worth It

This myth about location sounds logical on the surface: More sun must mean better solar power production, right? In reality, solar power system performance depends far more on total sunlight over the year than on how hot or sunny a place feels day to day. See Chapter 3 for more information about solar power and location.

Some of the strongest solar markets in the world are not especially hot. Germany, for example, has built massive amounts of solar capacity despite having a climate many people would describe as cloudy. The reason is simple: Panels respond to light, not heat, and they can perform very well in cooler temperatures. In fact, extreme heat can slightly reduce panel efficiency, meaning mild or cool regions often perform better than expected.

Another factor in looking at whether installing solar power is effective involves electricity pricing. Adding solar power production makes the most financial sense in locations where grid electricity is expensive or rising quickly, regardless of climate. A moderately sunny region with high utility rates can deliver better payback than a very sunny region with cheap electricity. That's why solar power adoption is growing rapidly in places like the Northeast and Midwest, not just the Southwest.

What really matters is combining a solar power resource with local economics and conditions. If a location has reasonable daylight, a suitable site, and significant electricity costs, solar power can work even if you don't live somewhere known for palm trees and constant sunshine.

Solar Panels Require Constant Maintenance or No Maintenance at All

The misconception about maintenance comes in two extremes. Some people imagine solar panels as delicate systems that need constant attention, while others believe they're completely foolproof, so set-it-and-forget-it. The reality sits comfortably in between.

Solar panels themselves have no moving parts, which makes them remarkably durable. Once installed, they quietly produce electricity day after day with very little intervention. In most cases, routine rainfall does a decent job of keeping panels clean, and modern systems are designed to withstand wind, snow, and temperature swings for decades.

That said, solar systems aren't entirely maintenance-free. Periodic checks help ensure that everything is working as expected. You should inspect system components — inverters, monitoring software, and electrical connections — occasionally, especially in the first few years. If a panel or inverter underperforms its warranty, monitoring systems usually catch it long before the homeowner notices a problem.

The key point is that solar maintenance is predictable and modest. It's nothing like maintaining a home furnace, power generator, or vehicle. You don't need weekly inspections, but ignoring the system entirely isn't wise either.

Solar Power Is Expensive

Installing a solar power system used to be expensive, and that history still colors how many people think about it today. Early systems were costly; incentives were limited; and payback periods were long. But evaluating modern solar power systems presents a very different scenario now. Consider that

>> **The cost of solar panels has dropped dramatically over the past decade.** This price drop (see Chapter 1) is driven by global manufacturing scale, improved efficiency, and streamlined installation practices. At the same time, financing options have expanded. Many homeowners and businesses no longer need to pay the full system cost upfront, and monthly solar payments can be comparable to, or lower than, a typical electric bill.

>> **The focus on upfront sticker price should sift to lifetime cost.** Installing a solar power system is an upfront investment that replaces decades of electricity purchases. When you spread that cost over 25 to 30 years of production, the effective price you pay per kilowatt-hour of energy use is often lower than grid power, especially as utility rates continue to rise.

In short, solar isn't cheap in the sense of being free. But for many users, it's one of the lowest-cost sources of electricity they'll ever buy.

Installing Solar Power Means You're Off-Grid

This off-grid misconception comes from confusing producing solar power with self-sufficiency. You can use solar power generation off-grid, but most solar systems today are connected to the electric grid and designed to work alongside it.

In a typical grid-connected setup, solar panels supply power to your home or business when the sun is shining. If the system produces more electricity than you're using at that moment, the excess flows back to the grid. When solar production drops, at night or during cloudy periods, electricity simply flows in from the grid as usual. From the user's perspective, nothing feels different. Lights stay on, appliances keep running, and reliability remains the same.

Going fully off-grid is a deliberate design choice that requires larger systems, significant battery storage, and careful energy management. See Chapter 9. It's common in remote locations where grid access is unavailable or extremely expensive, but it's not the default or the goal for most solar power adopters.

Solar Power Doesn't Work in the Winter or Snow

Winter is often seen as solar's weak season, but *weaker* doesn't mean *nonexistent*. Solar panels continue to generate electricity throughout the winter months, even in cold and snowy climates. Recognize that

>> **Cold temperatures actually help solar panels operate more efficiently.** While shorter days and lower sun angles reduce total electricity production when compared to summer, panels themselves perform better in cool conditions than in extreme heat. That's why winter output can be surprisingly strong on clear, cold days.

>> **Solar panels are typically installed at an angle.** The angle helps snow slide off as it melts, which makes the snow much less of a problem than many people expect. Dark panel surfaces also absorb sunlight and warm up quickly, accelerating snow shedding. Even when panels are temporarily covered, the impact is usually limited to short periods rather than entire seasons.

REMEMBER

The bigger picture matters most. Solar systems are designed based on annual energy production, not winter-only performance. Lower winter output is expected and accounted for during system sizing, just as higher summer production is part of the plan.

Panels Are Bad for the Environment

This bad-panel misconception usually comes from focusing on how solar panels are made rather than what they contribute over their lifetimes. It's true that manufacturing solar panels requires energy and raw materials, but that's true of every form of power generation, including fossil fuels.

What matters is the full lifecycle. A modern solar panel typically "pays back" the energy used to manufacture it within one to three years, depending on technology and location. After that, it produces clean electricity for decades. Over a 25- to 30-year lifespan, a panel avoids many times the emissions created during its production.

There's also growing attention on recycling and materials recovery. Early generations of panels weren't designed with end-of-life reuse in mind, but today's industry is moving quickly toward better recycling processes and circular supply chains. Valuable materials such as glass, aluminum, and silicon can already be recovered at high rates, and newer designs are improving even further.

Solar power systems aren't impact-free; no energy source or infrastructure is. But compared to extracting, transporting, and burning fuels year after year, solar power's environmental footprint is small, finite, and steadily shrinking.

Solar Can't Power Large Infrastructure or Industry

This misconception about the capabilities of solar power usually comes from picturing a solar system as something small — a few panels on a roof powering a house or a cabin. In reality, solar power systems already play a major role in powering some of the largest energy consumers in the world.

Utility-scale solar plants routinely generate hundreds of megawatts of electricity, enough to supply entire cities. Large corporations, data centers, manufacturing facilities, and even airports increasingly rely on solar power as a core part of their energy mix. These projects aren't symbolic; they're designed to offset meaningful portions of industrial load and reduce exposure to volatile energy markets.

Solar power works at scale because it's modular. Adding capacity doesn't require reinventing the system; it just means adding more panels, inverters, and land or rooftop space. When paired with storage, solar power generation can also deliver power during peak demand periods, support grid stability, and reduce reliance on fossil-fueled *peaker plants* (energy production plants that operate only when demand is at its peak).

The reality is that solar already does power large infrastructure. The better question today isn't whether solar can scale, but how quickly grids, markets, and policies can adapt to accommodate it.

There's Not Enough Land for All the Solar Power We Need

This concern about space for solar installations sounds reasonable at first glance, but it usually comes from underestimating how flexible solar siting can be and overestimating how much land solar actually requires.

Solar systems don't rely only on open fields or deserts. Rooftops, parking lots, warehouses, schools, factories, and *brownfield sites* (underutilized or abandoned properties) already provide enormous, untapped space. Many of these surfaces serve no energy purpose today, but they sit adjacent to where electricity is used. Using existing built environments dramatically reduces the need for new land altogether.

Even when ground-mounted solar structures are needed, the land footprint is often smaller than people expect. Studies consistently show that supplying a large share of electricity with solar power requires only a fraction of available agricultural or marginal land. In many cases, solar can coexist with farming through agrivoltaics, grazing, or pollinator-friendly ground cover.

Land use isn't a solar-specific problem; it's an energy-planning problem. When compared to mining, drilling, pipelines, and fuel transport infrastructure, solar systems' land impact is transparent, finite, and adaptable.

Solar Power Is Just a Fad

Calling solar power a fad assumes it's a temporary trend driven by hype rather than fundamentals. But solar power's growth isn't based on fashion or policy alone; it's rooted in physics, economics, and scale.

Scientists and engineers have studied the basic technology behind solar cells for more than a century, and commercial solar power has been deployed at meaningful scale for decades. What's changed recently isn't interest, but viability. Installation costs have fallen; efficiency has improved; and integration with grids and storage has matured. Those factors are structural shifts, not passing trends.

Solar power also benefits from something most fads lack: momentum that reinforces itself. As people deploy more solar power, manufacturing scales up, costs fall further, software improves, and workforce expertise grows. That feedback loop makes solar more competitive each year, not less. Utilities, corporations, and governments aren't betting on solar as an experiment, rather they're planning around it as long-term infrastructure.

Fads fade when conditions change. Solar persists because the conditions that support it — rising energy demand, the need for resilience, and the value of low-cost electricity — aren't going away. Solar power is here to stay.

Glossary

air mass: the thickness of the atmosphere that solar radiation must pass through to reach Earth's surface

alternating current (AC): electricity that periodically reverses direction. Homes, appliances, and the electric grid use AC power

asset management: oversight that sits above day-to-day operations and maintenance, including managing contracts, tracking warranties, overseeing compliance with financing agreements, coordinating insurance, and planning long-term upgrades or replacements

authority having jurisdiction (AHJ): the local government body that reviews and confirms a solar project meets building and electrical code requirements

autonomy: a design metric for off-grid systems representing how many days a home can run without solar input; a common target is three days of autonomy

azimuth: the compass direction your panels point, measured in degrees from north

backside generation: the ability of bifacial solar panels to collect reflected light from the ground or surrounding surfaces on the rear side of the panel

ballasted mounting: a commercial racking method that uses weight rather than roof penetrations to hold solar panels in place

bifacial panels: solar panels that can generate electricity from sunlight hitting both the front and back sides

brownfield sites: previously developed or underutilized land, such as former industrial properties, that can be repurposed for solar installations

by-right use: a zoning designation in some jurisdictions that allows solar projects to proceed on certain land types with minimal approvals

C&I: commercial and industrial; a category of solar installations on corporate campuses, retail stores, distribution centers, hospitals, and industrial facilities

cadmium telluride: a compound semiconductor known for excellent light absorption, used in thin-film solar panels

capacity factor: the percentage of a system's actual energy production compared to the maximum it could produce if it ran at full output all the time

capital stack: the combination of funding sources — such as equity, debt, and tax equity — that make a solar project financially viable

commissioning: the final phase of a solar project's construction arc, when the system is tested and verified to be operating correctly

curtailment: what happens when solar generation must be reduced because the grid cannot absorb all the electricity being produced

curtailment events: specific instances when a grid operator orders a solar plant to reduce output due to grid congestion or oversupply

demand charges: fees utilities charge commercial customers based on their highest 15-minute power spike in a billing cycle

direct current (DC): electricity that flows in a single, constant direction. Solar panels produce DC power

distributed generation: small-scale electricity production located near where energy is used, such as rooftop solar on homes and businesses

duck curve: a pattern in electricity demand in which demand remains high in the evening just as solar output declines, creating a curve that resembles the profile of a duck

electrification: the strategy of converting systems that previously ran on fossil fuels to run on electricity, often paired with solar and battery storage

energy: the total amount of electricity used or generated over time, typically measured in kilowatt-hours (kWh) or megawatt-hours (MWh)

Environmental, Social, and Governance (ESG): a framework companies use to evaluate sustainability performance and public commitments

escalators: contract provisions that allow payments to change over time based on triggers such as inflation or cost of living

evapotranspiration: the process by which plants release moisture, cooling the air around them; relevant to using vines and green walls to shade buildings

fixed-tilt system: a solar installation in which panels are mounted at a stationary angle rather than following the sun

force majeure: contract provisions in power purchase agreements that allocate risk for extraordinary events such as grid outages, natural disasters, or regulatory changes

Green Button initiative: a standardized system that allows utility customers to download their energy usage data in a common format

greenwashing: making misleading claims about the environmental benefits of a product or practice; solar energy's trackable metrics help counter greenwashing concerns

grid-tied setup: a solar power system configuration that remains connected to the utility grid, allowing two-way flow of electricity

ground-mounted system: a solar installation placed on racks anchored to the ground instead of on rooftops

heterojunction (HJT): a solar cell technology that layers amorphous silicon on crystalline silicon to improve efficiency through better passivation

independent power producers (IPPs): companies whose business model is to develop, own, and operate power plants and sell electricity under long-term contracts, rather than operating the grid itself

interconnection: the process by which a solar project applies to connect to the existing electrical grid and undergoes a series of technical studies to assess impacts

interconnection queues: the backlog of projects waiting for approval to connect to the electrical grid

inverter: a device that converts the direct current (DC) electricity produced by solar panels into alternating current (AC) electricity that homes, businesses, and the grid can use

investment tax credit (ITC): a U.S. federal tax credit that lets homeowners and businesses deduct a percentage of the cost of installing a solar energy system from their federal income taxes; as of this writing some of the credit is expired, and the remaining credit is scheduled to phase down and expire unless extended by Congress

irradiance: a measure of solar power received at a given surface area; design software pulls in historical irradiance data to estimate how much electricity a system will produce

kilowatt (kW): a unit of power equal to 1,000 watts, commonly used to describe system size

kilowatt-hour (kWh): a unit of energy equal to one kilowatt used for one hour; the unit utilities use to bill electricity consumption

levelized cost of energy (LCOE): a metric that calculates the average cost of producing electricity from a power system over its lifetime, including construction, operation, maintenance, and fuel (if any), divided by the total energy generated. Usually represented in \$/kWh or \$/MWh, LCOE allows apples-to-apples comparisons between different energy technologies

load patterns: the profile of how and when a building or facility uses electricity; sites with high daytime usage and stable load patterns are typically better solar candidates

load shifting: the practice of moving electricity consumption to different times of day to take advantage of lower rates or align with solar production

loan underwriting process: the process in which the lender thoroughly assesses the lending risk before approving financing for a solar installation

local jurisdictions: your city or county that governs building projects and adopts building and electrical codes

microgrid strategy: an approach in which a solar power system is combined with energy storage and other resources to create a self-sufficient local energy system, often used by hospitals and critical facilities

module (solar module): a single solar panel made up of interconnected photovoltaic (PV) cells that convert sunlight into electricity

multi-junction cells: solar cells that stack three or more materials to capture different parts of the solar spectrum; common in space applications in which efficiency outweighs cost

negawatts: the watts of energy you save instead of generate, through efficiency improvements and reduced consumption

net metering: a system of billing that credits the owners of solar power systems for excess electricity that their systems send to the grid

NOI multiples: a real estate valuation metric based on net operating income; solar can improve property valuations by lowering operational costs and increasing NOI

nuclear fusion: the process by which the Sun fuses hydrogen atoms into helium molecules, releasing enormous amounts of energy in the form of light and heat

off the grid: a solar power configuration in which a home or building operates independently from the utility company, generating and storing all its own electricity

offsetting emissions: reducing or eliminating the amount of pollution that would otherwise be created; every kilowatt-hour generated by solar power is one less that needs to come from fossil fuels

oversizing: intentionally installing more solar and battery capacity than typical usage requires, necessary in off-grid systems to account for worst-case conditions

passivation: a process in which thin layers of amorphous silicon coat a crystalline silicon surface, smoothing out defects and allowing electrons to move more freely, resulting in higher efficiency

peaker plants: power plants built to run only during periods of high electricity demand; solar paired with storage can reduce reliance on these fossil-fueled facilities

Permission to Operate (PTO): the formal authorization from a utility that allows a newly installed solar system to begin generating and exporting electricity to the grid

perovskites: a class of crystalline structures that are especially good at absorbing light and moving electric charge, used in emerging solar cell technologies

photon: a tiny packet of energy traveling from the Sun at the speed of light; the basic particle of light

photovoltaic (PV): a type of solar technology that converts sunlight directly into electricity using semiconductor materials

power: the rate at which electricity is produced or used at a given moment, typically measured in kilowatts (kW), megawatts (MW), or gigawatts (GW)

power purchase agreement (PPA): a long-term contract in which a buyer agrees to purchase electricity from a solar project at a predetermined rate

price cannibalization: a phenomenon in which wholesale electricity prices drop sharply during sunny hours in regions with high solar penetration, reducing the value of solar generation

production-based incentives (PBIs): programs that reward solar system owners for every kilowatt-hour their system produces

ratchets: utility rate provisions that lock in high demand charges for a full year based on a single peak usage event

rectennas: short for rectifying antennas; large receiving stations on the ground that convert beamed microwave energy from space-based solar into usable electricity

renewable energy: energy from sources that are naturally replenished, such as sunlight, wind, or flowing water

revenue optimization: the practice of maximizing what a solar project earns within its contractual and grid constraints, including verifying energy payments and optimizing inverter settings

satellite constellations: coordinated groups of orbiting satellites used in space-based solar power concepts to collect and beam energy to Earth

self-consumption: using as much as possible of your solar power production yourself, rather than exporting it to the grid

simple payback: the number of years until cumulative savings from a solar system equal the net cost of the system

single-junction: a conventional solar cell design using one layer of semiconductor material; tandem and multi-junction designs aim to push efficiencies beyond single-junction limits

solar carports: parking lot solar canopies that generate electricity while shading vehicles, turning single-use space into dual-purpose infrastructure

solar radiation: the full range of electromagnetic energy emitted by the Sun, including visible light, ultraviolet, and infrared wavelengths

solar tracker: a structure that rotates solar panels to follow the Sun throughout the day, increasing energy production but adding cost and mechanical complexity

string configurations: the arrangement of how solar panels are wired together and connected to inverters, modeled by design software to ensure energy flows efficiently

sun chart: a diagram that maps the Sun's movement and intensity across the sky for a given location, used in solar energy planning

sunlight: the visible portion of solar radiation — the familiar bright light that illuminates our days

supporting cast: the other system components beyond solar panels — including inverters, wiring, support structures, and batteries — that determine overall system performance

sustainable energy: energy from sources that are renewable (can be replenished naturally) and have minimal environmental impact, such as solar, wind, and hydro power

tilt angle: how steeply solar panels are slanted relative to the ground; the optimal angle for maximum annual production is typically close to the site's latitude

time-of-use (TOU): a utility rate structure in which electricity costs vary by time of day, with higher prices during peak demand hours

tracker: a mounting system that automatically moves solar panels to follow the sun's path across the sky, increasing energy production compared to fixed-tilt installations

utility-scale: large solar power installations that sell electricity into the grid, often owned by independent power producers or utilities

virtual power plant: a network of many small, distributed energy systems (such as solar-equipped homes with batteries) that collectively respond to grid conditions, functioning like a single large power plant

Index

benefits

  carbon emissions reduction, 265

  energy independence, 266–267

  energy price stability, 264

  ESG scores and public perception, 267–268

  job creation, 268

  local economic growth, 268

  lower electricity costs, 263–264

  modularity and scalability, 269

  property and asset value, 266

  with PV Systems, 52

  silent and low-maintenance operation, 268–269

  tax incentives and rebates, 264–265

branding opportunities, solar installation, 155–156

budgeting, long-term, 138

bureaucracy, 17

business type, solar installation, 135

buyers, 200, 202–204

  emerging, 203–204

  original, 203

  payment, 200

  request-for-proposals (RFP) process, 201

  types, 202

  utility-scale solar power, 202

C

cadmium telluride, 45

capability, solar, 145

  commercial properties, 146

  misconceptions, 276

  open ground, 146, 150–151

  parking lots, 146, 148–150

  rooftops, 146–148

capital stack, 194

carbon dioxide (CO_2) offset, 8–9

carbon emissions reduction, 265

carbon footprint, 8–9

carbon reduction, 139

carports, 148, 149

  supporting structures, 151

clothes dryer, 75

commercial and industrial (C&I) properties, 139

  common use cases, 140

  form and location, 140–141

  operational characteristics, 141

commercial solar power, 157

  annual energy production, 166

  buy-in, 163–164

About the Author

Ally Stone (DeGunther) With over a decade of experience in the renewable energy industry, Ally has been instrumental in advancing groundbreaking solar and storage solutions, contributing to two startups that each achieved $1B+ valuations. Ally has contributed to the development of gigawatts of utility-scale solar projects globally and has raised $1B+. She studied Mathematics at UCLA, which has equipped her with a strong analytical foundation that supports her strategic planning, data synthesis, and operational expertise. Her approach combines analytical rigor with a deep commitment to sustainable innovation, making her a driving force in the renewable energy industry. Outside of work, Ally enjoys exploring new technologies and is an avid cook.

Dedication

To my incredible husband, Micah — my constant light. Without him, the sun would never shine and none of this would be possible.

To my son, Calvin, we cannot wait to welcome you into the world. From the very first writing sample to the final page, I carried you with me. This is only the first of many things we will create together.

Author's Acknowledgments

This project wouldn't have been possible without my village. I'd like to thank my family — my brother and best friend Erik DeGunther and his partner Kristen Burkhart — who provide me with endless love and support to pursue my Kuleana. I'd also like to thank my in-laws, Diana Stone and Jim Larus, for celebrating me and this project with so much enthusiasm and encouragement. Thank you as well to Nicholas Larus-Stone and Katherine Mann, Jeremy, Gabi, Asher, and Blake Larus. I'd like to thank my aunt, Connie Cowan, who has encouraged me at every step of life.

My friends, who are basically family to me, give me endless energy and encouragement to do great things. Thank you to Dr. Jen Nebor, Sophia Shaheen, Adi and Tal Dahan, Brooke and Matthew Dina, Jess Fischman, and Hunter Kroll.

I'd like to thank Ben Nowack and Tristan Semmelhack at Reflect Orbital for having the courage to dream big and help pave the way for the future. Thank you as well to all my coworkers at Reflect who have been infinitely supportive of this project and my career in so many ways — you are the best team I could ask for.

None of this would be possible without the incredible mentors who have given me so much. Stephanie and Paul Perry, thank you for the endless guidance and encouragement. Martin and Karin Hermann, thank you for the opportunities of a lifetime and for helping pave the way. You both took a bet on me and it changed my life forever. Dr. Tom Buttgenbach and Dr. Finbar Sheehy, thank you for teaching me so much about finance and modeling. Denita Willoughbuy, who has always reminded me what I am capable of and I am the "prize." Debbie Builder, who is a world-class solar developer and dear friend. Manhal Aboudi, who was kind enough to teach me so much so early in my career. I'd also like to thank my teachers throughout my education who shaped me long before my career began — especially my high school math teacher, Carol Spiess, who always encouraged me to be myself and to love learning for the sake of learning.

Thank you to my parents, who are no longer here but whom I know are still with me. To my mom, Katie — you are the most loving person, and I thank you for encouraging me to be exactly who I am. To my dad, Rik, who started this book and passed away a few years after it was published, I'm grateful we could connect over solar power among golf and general physics discussions. Rewriting this has made me feel close to you in ways I never expected.

Thank you to my development editor, Leah Michael, for helping make this book what it is and for your incredible patience. Thank you Kelly Dobbs Henthorne my copy editor for improving my writing and Stephanie Perry my technical editor and dear mentor for helping ensure we provide the best and most accurate data. And thank you to associate acquisitions editor Elizabeth Stilwell at Wiley for this opportunity and support.

Publisher's Acknowledgments

Associate Acquisitions Editor: Elizabeth Stilwell

Development Editor: Leah Michael

Copy Editor: Kelly Dobbs Henthorne

Editorial Assistant: Nina Hook

Technical Editor: Stephanie Perry

Managing Editors: Sofia Malik and Murari Mukundan

Production Editor: Magesh Elangovan

Cover Image: © Martin Bröckel/ Wirestock Creators/stock.adobe.com